Science and Technology Policy for Sustainable Development

Science and Technology Policy for Sustainable Development

— *Editors* —

Tan Kha Sheng
Dragan Soljan

CENTRE FOR SCIENCE & TECHNOLOGY OF THE NON-ALIGNED AND OTHER DEVELOPING COUNTRIES (NAM S&T CENTRE)

2012
DAYA PUBLISHING HOUSE®
New Delhi - 110 002

ISBN 9789351241744

Centre for Science and Technology of the Non-Aligned and Other Developing Countries (NAM S&T Centre)
Core-6A, 2nd Floor, India Habitat Centre, Lodhi Road,
New Delhi-110 003 (India)
Phone: +91-11-24644974, 24645134, Fax: +91-11-24644973
E-mail: namstct@gmail.com
Website: www.namstct.org

Published by : **Daya Publishing House®**
A Division of
Astral International Pvt. Ltd.
– ISO 9001:2008 Certified Company –
4760-61/23, Ansari Road, Darya Ganj,
New Delhi - 110 002
Phone: 23245578, 23244987
Fax: (011) 23260116
e-mail : dayabooks@vsnl.com website :
www.dayabooks.com

Laser Typesetting : **Classic Computer Services**
Delhi - 110 035

Printed at : **Chawla Offset Printers**
Delhi - 110 052

PRINTED IN INDIA

PENASIHAT SAINS
KEPADAY.A.B. PERDANA MENTERI
ARAS 3, BLOK BARAT, BANGUNAN PERDANA PUTRA
PUSAT PENTADBIRAN KERAJAAN PERSEKUTUAN
62502 PUTRAJAYA, MALAYSIA
Tel.: 03-8888 4845
Faks: 03-8888 3437

Foreword

Science and Technology (S&T) policies for sustainable development in recent times have come to play a central role for continued and sustained national socio-economic development and growth, particularly in context of the NAM and other developing countries. In order to judiciously exploit their limited natural and financial resources and at the same time making an innovative environment to be able to withstand the international technological competition in various fields, many of these countries are struggling to even formulate appropriate S&T policies. For this they require much needed information and data banks, instruments and expertise, which particularly relate to S&T input data on R&D statistics,

I am happy to learn that NAM S&T Centre has been constantly organising international workshops, roundtables and training courses, numbering 82 till date, and has published 50 books in diverse areas of science and technology including policy formulation and implementation, sustainable environment, technology transfer, renewable energy sources, rainwater harvesting, information technology, nanotechnology, Intellectual Property Rights, small & medium enterprises, etc. The Centre also facilitates the scientists and scientific institutions to establish closer links in various fields of science and technology, for example, through its Fellowships schemes.

Among its activities, the NAM S&T Centre had organised an international workshop on "Science and Technology Policy for Sustainable Development" at Tehran, Iran during 5-7 January 2011 jointly with the National Research Institute for Science Policy (NRISP), Ministry of Science. Research and Technology of Iran. The workshop provided an academic platform to discuss the aforementioned issues related to formulation and implementation of S&T policies and developing a national innovation ecosystem for sustainable development. The present book is an outcome of the deliberations of this workshop and comprises 19 scientific and technical papers from the experts and professionals of 9 countries.

I compliment the NAM S&T Centre, particularly its Director, Prof. Arun Kulshreshta, for this initiative and bringing out this book for wider dissemination and exchange of S&T policy information among the members of the scientific community and the national policy makers in developing countries' I trust that the book will be a relevant step forward in exchange of information on S&T policy making for sustainable development and help promote South-South Cooperation.

Sincerely,

Zakri Abdul Hamid, ***PhD***

Science Adviser to the Prime Minister of Malaysia

Preface

In order to stimulate the process of economic development by appropriate use of their S&T resources, developing countries need to have in place a strong S&T policy. However, developing countries lack human resources with adequate expertise in S&T policy formulation and implementation. The political masters in developing countries need more and better information for the formulation, modification and implementation of their S&T polices in order to ensure a sustainable growth of their economy. S&T can play a central role in raising the standard of life of the people by, for example, creating wealth for all, making the country globally competitive, utilising natural resources in a sustainable manner, protecting the environment and ensuring national security.

In order to deliberate on these issues, the Centre for Science and Technology of the Non-Aligned and Other Developing Countries (NAM S&T Centre) jointly with National Research Institute for Science Policy (NRISP), Ministry of Science, Research and Technology of Iran organised a 3-days International Workshop on "Science and Technology Policy for Sustainable Development" at Tehran, Iran, during 5th–7th January 2011.

Participants came from 14 countries, including, Brunei, Cambodia, Croatia, India, Indonesia, Iraq, Kenya, Malawi, Oman, Pakistan, Sri Lanka, Sudan, Zambia and the host country Iran, and from the NAM S&T Centre, showing the importance of Science and Technology Policy and Sustainable Development in the developing countries.

The Workshop was conducted at the National Library of Iran at Tehran in seven technical sessions. The presentations at these technical sessions are summarized in a total of 20 papers. The papers covered a rather wide topic ranging from S&T policy to technology know-how. The intense deliberations from these presentations were in an open and friendly atmosphere.

Papers presented are of relevance to developing countries who aspire to develop, modify or amend their policies on science and technology taking into account environmental and sustainability issues.

In the concluding session, a set of recommendations, appended here, entitled 'Tehran Resolution on S&T Policy for Sustainable Development' was adopted after considerable deliberations and debate.

This book represents efforts of the NAM S&T Centre, its member countries and other developing countries. Particularly reference is made to the host country, Iran, also a member of the NAM S&T Centre, which took painstaking efforts in making this workshop a meaningful one. We acknowledge the NAM S&T Centre and the team from Iran for effective organisation of this workshop. Thanks are due to all the authors for timely submission and revision of their respective papers. Our special thanks are to Mr. Bandopadhyay of the NAM S&T Centre for detailing the arrangements for the Workshop and Dr. V.P. Kharbhanda of the NAM S&T Centre for seeing through the compilation and printing process.

Tan Kha Sheng, ***PhD***
Dean,
Research and Graduate Studies Office, Universiti Brunei Darussalam,
BRUNEI DARUSSALAM
Email: khasheng.tan@ubd.edu.bn

Dragan Soljan, MSEE
Director, PMU Science and Technology Project,
ZAGREB, CROATIA
Email: dragan.soljan@bicro.hr

Introduction

Science and Technology (S&T) is increasingly being recognised as the determinant factor of national economic development and growth in the developing countries. Most of these countries however still suffer from the absence of a meaningful S&T policy for stimulating the process of economic development for appropriate use of the S&T resources. They also lack suitable human resources with adequate expertise in S&T policy formulation and implementation. The decision makers in developing countries are in need of more and better information for the formulation and implementation of S&T polices including data on S&T statistics, human resources devoted to R&D, education of R&D personnel, measurement of technological innovation and S&T outputs like publications, patents, technologies developed and commercialised etc so that the desired modifications in the present policies can be incorporated in order to ensure the sustainable growth of the countries without compromising on their present resources and keeping them prepared to face the future challenges. These countries have to recognise the central role played by S&T in raising the quality of life of the people, particularly (a) by creating wealth for all, (b) by making the country globally competitive, (c) by utilising natural resources in a sustainable manner and (d) in protecting the environment and ensuring national security. The national S&T policies and their implementation should adjust to the new global realities and also satisfy social and economic goals in order to making them sustainable. In order to rectify the situation, the S&T policy in the developing countries needs to be formulated, or redefined as the case be, through consultations between the governments, institutions and industries by taking into account the specific technological needs and resources of the country.

In order to deliberate on these issues, the Centre for Science and Technology of the Non-Aligned and Other Developing Countries (NAM S&T Centre) jointly with National Research Institute for Science Policy (NRISP), Ministry of Science, Research and Technology of Iran organised a 3-days International Workshop on 'Science and Technology Policy for Sustainable Development' at Tehran, Iran during 5th–7th January 2011.

The Inaugural Ceremony of the workshop commenced with the traditional recitation of Holy Quran, and Hymn which was followed by the welcome address by the Secretary of the Workshop, Dr. Mohammad Hassanzadeh, Faculty Member, Tarbiat Modares University and Director, Department of Science Popularization at the National Research Institute of Science Policy (NRISP), Iran. After the introductory address and background of the Workshop presented by Prof. Arun P. Kulshreshtha, Director, NAM S&T Centre; Dr. Seyed Hassan Ghodsypour, President of NRISP; Dr. Mohammad-Javad Larijani, President of the Institute for Studies in Theoretical Physics and Mathematics (IPM); Mr. Esfandiar Ekhtiari, the Zoroastrian M.P. Education and Research Commission; and Dr. Mohammad Mehdi Nejad Nouri, Deputy Minister for Science, Research and Technology of Iran gave their individual remarks.

The workshop was attended by 237 senior professionals and research workers of 14 countries, including Brunei, Cambodia, Croatia, India, Indonesia, Iraq, Kenya, Malawi, Oman, Pakistan, Sri Lanka, Sudan, Zambia and the host country Iran (216 Participants), and from the NAM S&T Centre. The overseas participants were from Brunei [Prof. Tan Kha Sheng, Dean of Graduate Studies and Research and Dr. Saiful Azmi Husain, Lecturer, the University of Brunei Darussalam, and Dr. Pg Mohd Esa Al-Islam Pg Hj Md Yunus, Acting Deputy Dean, Postgraduate Studies and Research and Dr. Pg Saiful Baharun Pg Duraman, Lecturer, Faculty of Engineering in Institut Teknologi Brunei, Gadong]; Cambodia [Dr. Sopheap Seng, Head of Department, Institute of Technology of Cambodia, Phnom Penh]; Croatia [Prof. Dragan Šoljan, Director, S&T Project Croatia, Ministry of Science Education and Sports, Zagreb]; India [Prof. (Dr.) Palpu Pushpangadan, Director General, Amity Institute for Herbal and Biotech Products Development, Kerala; Prof.. Dr. Parmod Kanwar Gupta, Secretary General and Principal Management Consultant, National Foundation of Indian Engineers (NAFEN), New Delhi; and Dr. (Mrs.) Purnima Rupal, Scientist E-II, International S&T Affairs Directorate, Council of Scientific and Industrial Research (CSIR), New Delhi]; Indonesia [Prof. Dr. Otto S. R. Ongkosongo, Senior Research Professor on Coastal Environment Research Centre for Oceanography, Indonesian Institute of Sciences (LIPI), Jakarta]; Iraq [Mr. Hadi Obaid Ali, Director General and Dr. Sabah Azeez Hasan, Deputy Director General, Ministry of Science and Technology, Baghdad]; Kenya [Mr. Frank Ndaks Sawanga Ndakala, Directorate of Research Management and Development, Ministry of Higher Education, Science and Technology, Nairobi]; Malawi [Mr. Patrick J.J. Mphadzula, Principal Science and Technology Officer, Ministry of Education, Science and Technology, Department of Science and Technology, Lilongwe]; Oman [Mr. Khalid Ali Salim, Financial Analyst, Ministry of National Economy, Muscat]; Pakistan [Mr. Muhammad Zafar Iqbal, Deputy Scientific Advisor, Ministry of Science and Technology, Islamabad]; Sri Lanka [Prof. Nalini Beatrice Ratnasiri, Chairperson, National Science and Technology Commission (NASTEC), Colombo]; Sudan [Prof Dr. El Tayeb Edris Eisa Ebrahim, President, Sudan Academy of Sciences, Khartoum]; and Zambia [Mr. Ernest Peter Shamano, Technology Audit and Validation Specialist, National Technology Business Centre, Lusaka]. From the NAM S&T Centre, Mr. M. Bandyopadhyay, Senior Expert and Administrative Officer and Prof. Arun P. Kulshreshtha, Director attended the event.

The overall programme of the Workshop was conducted at the National Library of Iran at Tehran in seven technical sessions. These sessions were respectively co-chaired by (*i*) Prof. Nalini Ratnasiri of Sri Lanka and Mr. Dragan Šoljan of Croatia; (*ii*) Prof Dr. El Tayeb Edris Eisa Ebrahim of Sudan and Prof. Dr. Jafar Mehrad of Iran; (*iii*) Prof. Dr. P. Pushpangadan of India and Prof. Tan Kha Sheng of Brunei; (*iv*) Dr. Pg Mohd Esa Al-Islam Yunus of Brunei and Mr. M. Bandyopadhyay of the NAM S&T Centre; (*v*) Prof. Dr. Arun P. Kulshreshtha of the NAM S&T Centre and Mr. Muhammad Zafar Iqbal of Pakistan; (*vi*) Dr. M. Molanejad of Iran and Mr. Hadi Obaid Ali of Iraq; and (*vii*) Dr. Ernest Peter Shamano of Zambia and Prof. P. K. Gupta of India.

The scientific papers presented during the Workshop were on 'SC: A New Bibliometric Citation System Applied to Research Journals in Islamic Countries' by Prof. Ja'far Mehrad (President, Islamic World Science Citation Center, Iran); 'Foresight of the Most Appropriate Technologies for Iran 2025 – Pilot Stage–(PAMFA)' by Dr. Aria Alasty (NRISP, Iran); 'SEMAT, National Current Research Information System for Iran' by Dr. Omid Fatemi (IRANDOC, Iran); 'Technologies for Sustainable Development' by Prof. Dr. El Tayeb Idris Eisa (Sudan); 'Keys to Success for the Sustainable Development of Modeling and Simulation Research Cluster' by Dr. Saiful Azmi Hussain (Brunei); 'How to Prepare the Action Plan for Implementing the National Science and Technology Policy' by Mr. Dragan Šoljan (Croatia); 'The Role of Science and Technology in Biodiversity Based Sustainable Development' by Dr. P. Pushpangadan (India); 'Forging International S&T Cooperation for Sustainable Development: Lessons from CSIR' by Dr. Purnima Rupal (India); 'Science and Technology Policy and Sustainable Development' by Mr. Ernest Peter Shamano (Zambia); 'Developing Science and Technology Research in the Institute of Technology of Cambodia' by Dr. Sopheap Seng (Cambodia); 'Energy Efficiency and Conservation for Sustainable Development: a Brunei Darussalam and Institute Technology Brunei Perspective' by Dr. Pg Saiful Baharun Pg Duraman (Brunei); 'The National Science and Technology Policy of Sri Lanka' by Prof. Nalini Beatrice Ratnasiri (Sri Lanka); 'CO_2 Capture and Sustainable Development of Micro-Algae Based Fuel' by Prof. Nasrin Moazami (Director, Institute of Advanced Technology, Iran); 'S&T Statistics of Pakistan: a Country Report' by Mr. Muhammadd Zafar Iqbal (Pakistan); 'Overview of the Science and Technology Institutional and Regulatory System in Malawi' by Mr. Patrick J. J. Mphadzula (Malawi); 'Evolution of Science and Technology Policy for Developing an Innovation Ecosystem and Promotion of Knowledge-Based Industries in India' by Mr. M. Bandyopadhyay (India); 'Towards Science and Technology Policy in Oman' by Mr. Khalid Ali Salim Al Senadi (Oman); 'Science and Technology in Brunei Darussalam' by Dr. Pg Mohd Esa Al-Islam Yunus (Brunei); 'The 50 Principles for Sustainable Development' by Prof. Dr. Otto. S. R. Ongkosongo (Indonesia); 'Women Empowerment through Science and Technology Education' by Dr. P. K. Gupta (India); 'University Brunei Darussalam Research Strategies and Management' by Prof. Tan Kha Sheng (Brunei); 'The Impact of Educational Tourism on Sustainable Economic Development' by Dr. Mohammad Hossein Karim, Science and Education Director of Iran in India and Ms. Malihe Esmaeili, Ph.D. Scholar, University of Pune (India), presented by Ms. Esmaili; 'Using Stakeholder Engagement to Bridge the Gap Between Science, Technology and Innovation (STI) Policies in

Implementing the Kenya Vision 2030 for National Development' by Mr. Frank Ndaks Sawanga Ndakala (Kenya); 'The Role of Science and Technology in Managerial Development' by Dr. Sabah Aziz Hassan and Mr. Hadi Obaid Ali (Iraq) presented by Dr. Hassan; 'Science and Technology Park – Bridge between University and Industry' by Assoc. Prof. Dr. Mahmood Torabi-Angaji (Adviser to Dean on International Affairs, Tehran University, Iran); 'Public Understanding of Science and Technology: a National Prototype for Popularization of Science, Policy and Actions in Islamic Republic of Iran' by Dr. Mohammad Hassanzadeh (Tarbiat Modares University, Iran) and Ms. Azadeh Heidari (Islamic Azad University, Iran), presented by Ms. Heidari; 'Science and Technology Policy in Islamic Republic of Iran' by Dr. Amir Ali Seifoddin (Office of S&T Development, Vice- presidency of Iran for Strategic Planning and Control) and Prof. M. Hasan Sadeghi Moghadam (Sharif University of Technology, Tehran, Iran), presented by Dr. Seifoddin; 'Role of the NAM S&T Centre for South – South Cooperation in Science and Technology' by Prof. Arun P. Kulshreshtha (NAM S&T Centre); 'Chemical Coatings' by Mr. Ali Mohamad; 'Solar Cells Unmanned Aerial Vehicles' by Mr. Milad Hajikhan; 'Autopilot Hovercraft' by Mr. Adel Kerami; 'Gold from Anodic Copper' by Mr. Reza Ranjbar; and 'Gas Nanosensors' by Ms. Nafees Hassanzadeh, all from Amir Kabir University of Technology, Tehran, Iran.

A special Participants' Forum was also organised during the Workshop, which was moderated by the Director, NAM S&T Centre. The discussion was highly interactive in nature and specific suggestions and comments were invited from the participants.

In the concluding session, a set of recommendations entitled 'Tehran Resolution on S&T Policy and Sustainable Development' was adopted after considerable deliberations and debate.

The present book is mainly based on the deliberations of the Tehran Workshop. I acknowledge with gratitude the deep involvement and determined efforts of Prof. Tan Kha Sheng, Dean of Graduate Studies and Research, University of Brunei Darussalam and Dr. Dragan Soljan, Director, S&T Project Croatia, Ministry of Science Education and Sports of Croatia in editing this valuable publication. I am also indebted to the entire team of the NAM S&T Centre, particularly Mr. M. Bandyopadhyay, Dr. V.P. Kharbanda, Mr. Yasir Abbas Rizvi, Dr. Gurjeet Kaur and Mr. Pankaj Buttan in compiling the presented papers, liaising with the authors and editors and giving a final shape to this volume.

I am confident that the materials presented herein would serve as a valuable reference material not only for the scientists, policy makers and executives, but also for the concerned authorities of the developing countries in taking appropriate policy measures.

Prof. Arun P. Kulshreshtha
Director and Executive Head,
NAM S&T Centre

Contents

Chapter 1

Universiti Brunei Darussalam Research Strategies and Management

Tan Kha Sheng
Universiti Brunei Darussalam
E-mail: khasheng.tan@ubd.edu.bn

ABSTRACT

The mission of Universiti Brunei Darussalam (UBD) is to develop the individual and society as a whole, guided by our commitment to the nation that we love and by extension, the world that we cherish. We will achieve this by pushing back the frontiers of human knowledge through the cultivation and enrichment of the intellect, faith and culture which will in turn be accomplished through the creation of a conducive environment for achieving excellence in teaching, learning, research, scholarship, public service and professional practices. In pursuing our mission, we shall be guided by the underlying values of our national philosophy. This paper will present the university's research strategies and management toward achieving the above mission.

Keywords: *UBD, ICube, Belalong Initiatives, KBFSC, IBM-UBD.*

Introduction

The idea of establishing the Universiti of Brunei Darussalam was first raised during a comprehensive review of the country's higher education facilities in 1976. Subsequent discussions led to the appointment of a steering committee which, with the assistance of the British Council, proceeded to engage a group of experts to study and advise on the matter. However, active planning for a university did not commence until 1984-5, at which time academic links were discussed with a number of universities in the United Kingdom and Malaysia.

On 23 April 1985 came an announcement from His Majesty Sultan Haji Hassanal Bolkiah Mu'izzaddin Waddaulah, Sultan and Yang Di-Pertuan of Negara Brunei Darussalam that, in the interest of national development, Brunei Darussalam was to have its own university and that the necessary arrangements should be made with minimum delay.

In a remarkably short period of time a temporary campus was set up in a renovated building complex close to the Institute of Education, four and a half kilometres from the city centre.

With the help of several overseas universities and under the supervision of Ministry of Education Committee on the Establishment of the University, the first degree programmes and courses were devised. Formal academic links were established with University College, Cardiff, and with the University of Leeds and both of these universities helped in the development of the English-medium programmes. For Malay-medium programmes, assistance was provided by Universiti Sains Malaysia and Universiti Kebangsaan Malaysia.

Academic staff were recruited in the United Kingdom, Malaysia and Singapore, and some were also seconded from universities in these countries.

On 28 October 1985, just six months after His Majesty's announcement, Universiti Brunei Darussalam (**UBD**) opened its doors to the first intake of 176 students. In 1988, the Sultan Hassanal Bolkiah Institute of Education was integrated with the University.

Objectives of UBD

UBD will, within the planning period aim to:

- ☆ Improve the quality of its graduates;
- ☆ Offer postgraduates opportunities in selected disciplines at the appropriate time and will at least double postgraduate student intakes;
- ☆ Give emphasis to applied research,
- ☆ Actively promote its expertise within the community by offering advisory, consultancy and continuing education services;
- ☆ Continue to improve the quality of its staff, in terms of qualifications and skills, and increase numbers to appropriate levels;
- ☆ Continue to improve the level of support facilities in order to ensure excellence in teaching, research and management;
- ☆ Respond effectively and appropriately to present and future needs and challenges.

UBD Strategic Goal in Research

The university's strategic goal in research is to contribute significantly to the development of pure and applied knowledge through various research programmes to be supported by both the university, the government and the business sector and jointly with external researchers and institutions. This goal is align with the Ministry of Education Strategic Plan (2006-2015), the National Development Plan (2006-2015) and the Prime Minister Office Strategic Plan (2006-2015).

The university's strategic research trusts are in Energy, Biodiversity and Bio-prospecting, Food security/Biotechnology and Asian Studies (Include Malay Islamic culture).

To this end various initiatives and activities are currently being carried out, *viz;*

1. The Kuala Belalong Field Studies Centre (KBFSC) was set up by UBD deep in the country's unspoiled jungle as an international focus for research into the threatened rainforests of Borneo. It is also a teaching and training resource for schools, institutes of higher education in Brunei Darussalam and overseas, and government personnel working on the environment. (Note: It is not an area for ecotourism or a recreation site).
2. Research area from national S&T funding (up to 2012) in Materials and Energy, Biodiversity and Bioprospecting, Medical, other S&T and Education.
3. Establishment of research clusters in Materials, Energy, Modelling and Simulation, Environment and Sustainable Development, Science Technology and Society, Ageing, Obesity, Cancer, Islamic Studies and Youth.
4. The Belalong initiatives that focus on biodiversity, alternative energy studies, food security, environment and sustainable development while at the same time addressing the social and economic needs of the country. List of projects include Climate Change and Biodiversity Studies, Education, Environment, Sustainability and Economy, Alternative Energy Sources for the KBFSC and Bio-prospecting and Natural Products.
5. The International Consortium of Universities for the Study of Biodiversity and the Environment (**ICUBE**). Membership in this consortium is made up of world class universities from the region, East Asia, Europe and the United States. The consortium will provide the framework for cooperation in research, teaching and learning on issues related to biodiversity, climate change and the environment. So far members of ICUBE include University of North Carolina, USA, King's College London, UK, University of Auckland, NZ, Korea University, Korea, Monash University, Australia and Bonn University, Germany.
6. Establishment of UBD-IBM Centre. This centre will collaborate on climate modelling research. This will investigate the impact of climate change on flood forecasting, crop yields, renewable energy and the health of rainforests in the region. The collaboration will help UBD accelerate its research capabilities in biodiversity, energy and agrotechnology. An IBM Blue Gene supercomputer is set up for these purposes.

Conclusion

Universiti Brunei Darussalam (UBD) aims to produce quality graduates suitably equipped in terms of knowledge, skills, attitudes, moral and spiritual values, to support the development needs of the nation and consistent with the national philosophy. UBD also promotes and undertakes research, particularly applied research, in areas where it has a comparative advantage and in accordance with national needs.

Today the University accommodates more than 350 academic staff and 3,600 students in 7 faculties and strives to serve the Nation in many different fields of application.

From August 2009, UBD has a new generation of undergraduate and postgraduate degrees. This change will align the University with the best of Asian and European practices. In reviewing its programmes, UBD has been inspired by His Majesty's Titah given during the recent UBD 20th Convocation ceremony in which he urged the University to offer programmes that are more innovative, broad-based and multi-disciplinary. As the premier institution of learning, UBD needs to take up the mantle of being the pioneer among the higher education institutions in the country, with a strongly evident competitive edge.

Acknowledgments

The author would like to thank His Majesty the Government of Brunei Darussalam and the NAM S &T Centre for their kind support.

References

Brunei National Development Plan (2006-2015).

Ministry of Education Strategic Plan (2006-2015).

The Prime Minister Office Strategic Plan, Brunei Darussalam (2006-2015).

Universiti Brunei revised corporate strategy map, 2009.

Chapter 2

How to Prepare the Action Plan for Implementing the National Science and Technology Policy

Dragan Soljan[1] & Damir Vucina[2]

[1]MSEE, Director Science and Technology Project,
Ministry of Science, Education and Sports of the Republic of Croatia
E-mail: dragan.soljan@bicro.hr
[2]Faculty of Electrical Engineering,
Mechanical Engineering and Naval Architecture, Split, Croatia
E-mail: vucina@fesb.hr

ABSTRACT

This paper examines the question of what happens following the creation of a national Science and Technology (S&T) Policy and presents a "cookbook" approach to preparing the Action Plan for implementing the Science and Technology Policy, based on experience gained in preparing the Action Plan accepted by the Croatian Government covering the period 2007-2010. Following the S&T Policy, a document clearly assigning goals, responsibilities, priority areas, proposed actions and progress indicators is essential to fully follow through the implementation. The lack of such a document would most likely leave the S&T Policy on the shelf with other strategic documents in the library section of wishful thinking.

We propose to commence with a brief overview of the Background and Rationale for such a document. This should be followed by a critical and focused Gap Analysis of the National Innovation System.

It is essential to carefully select and explicitly state the Progress Indicators for monitoring implementation and assessment of impacts of this Action Plan. The selected indicators need to be measurable and verifiable, as well as preferably available from public sources. Most importantly, they need to be relevant to the goals.

Define a list of proposed Actions for individual policy elements. Each action/policy element should state the implementing body, measures and instruments available, specific objectives, key outcomes envisaged, related milestones and relevant progress indicators.

The final section of the Action Plan may recognize priority areas, *i.e.* niches where your country has competitive advantages, or areas which you wish to develop in accordance with strategic priorities.

Obviously, there is not just one correct way to implement your S&T Policy, and successful outcomes are highly dependant on a complex set of external factors, often beyond our influence and control. This paper outlines the recommended structure of an Action Plan containing all the required ingredients.

Keywords: *Action plan, Science and technology policy, National innovation system, Gap analysis, Progress indicators, Priority areas.*

Introduction

There is plenty of empirical evidence that Science and Technology (S&T) are the key drivers of economic growth and there is a direct correlation between S&T utilization and economic competitiveness and overall social welfare. Innovation is the cornerstone of the European Union's (EU) 'Lisbon' strategy again emphasized by the EU 'Barcelona' conclusions. The EU has cited insufficient innovation capacity as a key factor for the EU's underperformance in productivity growth. S&T Policy must go beyond the model of Research and Development (R&D) as a source of innovation to more complex models of an interconnected system of mutually interacting institutions. The 'business-as-usual' increases the developmental gap between the wealthy countries (regions) and the underdeveloped ones. Developing countries in particular need to 'run faster' than developed countries in order to catch up, or even just not to fall behind.

Innovation may be triggered by R&D supply, by market demand, by necessary upgrades in competitiveness, by emerging of new companies, etc. The S&T Policy must design a friendly and supportive environment to encourage and maximize benefits. It must not only focus on technology creation but also on technology diffusion, adaptation and transfer into the business environment, since these mechanisms are more frequently the factors for development. Both incremental and radical innovations are fundamental drivers of economic development. In order to create synergies, the S&T and innovation policies must be interconnected with sector-specific and other policies.

This paper examines the question of what happens following the creation of a national Science and Technology Policy and presents a "cookbook" approach to preparing the Action Plan for implementing the Science and Technology Policy, based on experience gained in preparing the Action Plan accepted by the Croatian Government covering the period 2007-2010 (MSES, April 2007). Following the S&T Policy (MSES, October 2006), a document clearly assigning goals, responsibilities, priority areas, proposed actions and progress indicators is essential to fully follow through the implementation. What follows is a proposed structural outline for such a document.

How to Prepare the Action Plan

Background and Rationale

We propose to commence with a brief overview of the Background and Rationale for such a document. This introductory section should limit itself to re-stating the key goals of the existing S&T Policy, while highlighting the country specifics, thus firmly anchoring the document in the economic and scientific and technological environment it is growing on. Keep in mind wider regional demands which also influence your country.

Knowledge is transferred into goods and services through a country's National Innovation System (NIS), which encompasses a policy framework conducive for innovation, knowledge generating organizations (Higher Education Institutions–HEIs, and Public Research Organizations–PROs), an Intellectual Property Rights (IPR) regime, Measurements, Standards, Testing and Quality (MSTQ) systems, as well as a modern information access infrastructure. The objective of a well functioning NIS is to serve the needs of the economy by achieving a better integration of the technology and innovation infrastructure with market needs, increasing private sector participation, and developing linkages between industrial and research communities. There is an evident need for a greater contribution from the academic sector to sustainable regional development and industry competitiveness. The main obstacle to the transfer of this knowledge is an evident gap between the commercialization of academic research outcomes (supply side) and industry needs (demand side).

Within this context, the Government of the Republic of Croatia has moved to the implementation stage of its ambitious agenda in developing its NIS through leveraging highly skilled human capital, improving efficiency of the scientific research know-how present in the country, creating appropriate mechanisms to share the existing knowledge throughout society, and developing a conducive environment for upgrading companies' innovation capabilities in order to increase their competitiveness.

The Ministry of Science, Education and Sports (MSES) is the central authority in charge of implementing the National Innovation System, comprising activities for fostering the development of Croatia as a knowledge-based society with an economy based on innovation. The NIS Council (NISC) within the MSES acts as a "think-tank" assisting the Ministry in the design, analysis and implementation of various policies and programs to strengthen the Croatian NIS, fostering changes in the current production model into a model of sustainable development based on innovation (Table 2.1). The NIS supports financing innovations, from basic research in priority areas, fostering technology transfer from HEIs and PROs to the private sector, stimulating innovations in the private sector, and promoting venture capital in Croatia with the goal to stimulate industrial growth, increased productivity and exports, and improved positioning of the national economy in view of expected EU-accession in the near future.

There needs to be a systems approach in the restructuring of the Croatian S&T system, with a clear vision, objectives, objective-based incentives, programs and measures, targets; roles, instruments, monitoring indicators, which will close the existing gaps and resolve system deficiencies in a structured and functional way. The systems approach will create synergies within the S&T system and leverage much more impact than individual non-coordinated interventions.

The Croatian S&T and innovation policy must be compatible with that of the EU which emphasizes:

- ☆ Coherence of innovation policies
- ☆ Regulatory framework conductive to innovation
- ☆ Encouragement of creation and growth of innovative enterprises
- ☆ Improvement of the key interfaces in the innovation system
- ☆ Society open to innovation but also take into account the specifics of the Croatian situation.

Basic Gap Analysis of Current National Innovation System (NIS)

Only you know the country-specific problems and limitations afflicting the public and private sectors (and their complex interactions or lack thereof) in your country. Being brutally sincere at this point makes the following steps much easier.

Below is a list of elements considered in the case of the Croatian Action Plan (MSES, April 2007):

- ☆ Competitiveness of Croatian economy must grow to a satisfactory level,
- ☆ Major transitional structural changes, large loss of industrial capacity and exports, economy at-large running into low-tech and low-profit labor-intensive areas, trade dominates in economy,
- ☆ Low level of patent applications, low commercial exploitation,
- ☆ Croatian science and technology system should globally be more competitive and contribute more towards development,
- ☆ S&T policy and innovation policy not a major component in the overall development policy,
- ☆ Very little interaction between public S&T and economy (industries, SMEs), the interaction needs to evolve from R&D sponsorship to forms of mutual participation in common projects with common interest, which should also be supported by the regulatory framework,
- ☆ Low investment into R&D from the business sector,
- ☆ Internal R&D investments within companies limited and present only in few large companies,
- ☆ Lack of awareness towards commercially exploiting R&D results within the public RDIs and universities,
- ☆ Academic curricula not flexible,

- ☆ Academic curricula insufficiently correlated with the needs of the economy,
- ☆ Very little mobility, both within public S&T system and between public S&T and industry,
- ☆ Insufficient presence in international S&T cooperation,
- ☆ Insufficient direct commercial impact of the public S&T investments and funding,
- ☆ Insufficient and unfocused funding of S&T, results not evaluated or monitored properly,
- ☆ No satisfactory information system that would provide well-structured data available for evaluation, benchmarking and monitoring, no fully-adequate 'management information system' for the public S&T in place to capture their mutual impact, cost-benefit assessments, etc.,
- ☆ Very little motivation for S&T careers, brain-drain and brain-waste situation
- ☆ Decreasing enrollment and interest (in relative terms) for S&T careers,
- ☆ Promotion system in HEIs and PROs that essentially measures only the 'number of papers' and disregards many other S&T related individual achievements.

Aggregate Progress Indicators

It is essential to carefully select and explicitly state the progress indicators for monitoring the implementation and assessment of impacts of this action plan. The selected indicators need to be measurable and verifiable, as well as preferably available from public sources. Most importantly, they need to be relevant to the goals. Set a baseline and time-based, realistic but ambitious goals. Grouping your indicators under R&D supply and absorption-capacity demand-side indicators can help to increase clarity.

Some possible sources for aggregate progress indicators include:

- ☆ EU Innovation Scoreboard
- ☆ OECD S&T indicators
- ☆ WB competitiveness indicators

The following indicators were selected for the monitoring of implementation and assessment of impacts of the Croatian Action plan:

On the R&D Supply Side

T1. R&D expenditure as percentage of GDP

currently: 1.05 per cent (2004)

T2. Number of researchers (FT) per 1000 labor force

currently: 3.8 (2003)

T3. Number of patent applications to the EPO per million labor force

currently: 48.8 (2002)

T4. Number of CC/SCI publications per 1000 R&D employees

T5. Number of Science and Engineering (SE) graduates per 1000 population aged 20-29

currently: 5.4 (2004): women 3.6 (2004)

T6. Number of S&E PhDs per 1000 population

currently: 3.5

On the Absorption capacity and demand side:

T7. R&D and innovation expenditure of companies as percentage of GDP

currently: 0.52 per cent (2004)

T8. Number of employees participating in lifelong learning

T9. Number and value of R&D contracts between industry and HEIs/PROs

T10. Number of S&T start-up SMEs

T11. Percentage of total employees in high-tech service sector

currently: 3.29 (2005)

T12. Percentage of employees in high-tech manufacturing and industries

T13. Sales of new-to-market or new-to-company products as percentage of total sales

T14. High-tech exports – exports of high-technology products as a share of total exports

currently: 11 per cent (2004)

T15. Number and value of participation of Croatian researchers in international R&D projects

Targets were set for planned annual increases of all indicators. The targets were set optimistically high, to encourage growth and stimulate annual monitoring and corrections.

List of Proposed Actions for Individual Policy Elements

Define a list of proposed Actions for the individual policy elements. This list should be limited to the priority actions, and deduced from a much wider list of all possible actions envisaged by the S&T Policy (MSES, October 2006). Each action/ policy element should state the implementing body (agency, Ministry), measures and instruments available, specific objectives, key outcomes envisaged by the action, related milestones and relevant progress indicators (from the set defined previously).

The first recommended action is establishing an NIS Council (NISC) as an expert body in-charge of high-level coordination of actions aimed at operating the NIS. The NISC monitors and evaluates outcome of activities concerning all Policy Elements and revises this Action Plan on an annual basis.

Table 2.1: Establish National Innovation System Council (NISC)

Implementing agency: MSES, NISC

Measures, instruments: definition of mission, set-up, appointments establish expert groups, order expert services, monitor all developments in NIS, including implementation of this Action Plan.

Specific objectives:

- ☆ Design of objective-based coherent mix of policies and instruments.
- ☆ Evaluation and transfer of best-practices from abroad related to NIS design and programs, compatibility with EU schemes and regulations.
- ☆ Coordination of S&T policy measures with other sectors and actors, set up foresight activities.
- ☆ Development of new programs and measures (and upgrade of existing) based on gap analysis and identification of needs, in line with EC state aid rules.
- ☆ Development of the basic layout of the system for monitoring and benchmarking for elements of S&T network.
- ☆ Design and start the system of measuring inputs, outputs and impact of all NIS instruments.
- ☆ Continuous monitoring and upgrades of the NIS.
- ☆ Coordination of the development of the NIS information system, enhancing the flow of information.
- ☆ Introduce corresponding education and training at all levels.
- ☆ Reporting to MSES.
- ☆ Developing programs to reduce brain drain and expand international cooperation.
- ☆ Achieve full cycle in funding from setting priorities, competitive financing, strict evaluation, thorough monitoring, proper project management.

Key outcomes: design and redesign of several NIS components, more effective funding, leverage effects.

Milestones: establish by March 2007 individual milestones for different items annual update of S and T Action Plan.

Progress indicators: Mean value and variance of all targets set for indicators T1- T15. In case of high variance, selectively reinforce activities targeting at less performing indicators or make additional provisions in corresponding measures.

In order to secure political support a high-level sponsor is mandatory. The Governmental Body for Science and Technology (GOVBOST) is the high-level sponsor for the implementation of Science and Technology Action Plan, providing approval for all activities, and Government support for implementation. The GOVBOST in Croatia was headed by the Prime Minister, and included the Minister of Science, Education and Sports, Minister of Economy, Labor and Entrepreneurship, Minister of Finance, Minister of Sea, Tourism, Transport and Development, Minister of Environmental Protection, Physical Planning and Construction and two members representing the R&D community (Table 2.2).

The priority actions are heavily country-dependant, but in general encompass: Improving the Legislative and Regulatory Framework, Educating the S&T Community towards Commercialization of R&D, Improving Financing of S&T, Intensifying R&D in the Private Sector, Integration with the International Environment as well as a Publicity Campaign aimed at Raising General Awareness on the Importance of S&T, etc. (Table 2.3).

Table 2.2: Establish Governmental Body for Science and Technology Policy (GOVBOST)

Implementing agency: Governmental body headed by Prime Minister (GOVBOST)

Measures, instruments: policy directions and approval, overall coordination, providing Government support

Specific objectives: create necessary support for implementation of S&T Action Plan, provide support for regulatory changes

Key outcomes: Active support of S&T Action Plan implementation, delegated individual responsibilities

Milestones: established March, 2007

Progress indicators: Action plan implementation underway, responsibilities delegated, budget approved

Table 2.3: Improve Legislative and Regulatory Framework

Implementing agency: Coordinated action of NISC and GOVBOST. Other participating agencies: respective Ministries, Croatian Accreditation Agency (CAA), State Intellectual property office (SIPO) and other government agencies.

Measures, instruments: proposals for changes in applicable Laws, support in obtaining approval in Parliament.

Specific objectives:

Diagnose deficiencies and obstacles in regulatory framework

- IP,
- taxation,
- accounting standards,
- S&T system promotion rules,
- HR and reward policy,
- Change promotion and reward legislation, policies and rules in the public R&D system to account for relevant forms of R&D excellence.

Key outcomes: changes in regulations, more supportive regulatory framework

Milestones: during 2007 update Law on Science and Higher Education.

- Other legislative changes to follow

Progress indicators:

1. necessary changes identified
2. changes proposed and coordinated
3. changes in regulations passed

In developing countries, the S&T sector is traditionally concentrated in the public domain, encompassing the national Higher Education Institutions (HEIs) and Public Research Organizations (PROs). Enhancement in this area should start upstream by addressing the following issues:

- Basic versus applied research
- Academic freedom versus responsibility to the investors (tax payers)
- Entrepreneurial climate and risk taking culture
- Publishing versus industry sponsored or joint projects
- IP portfolio management

Two main external factors present risks to the achievement of the above formulated objectives: a renewed resistance to the commercialization of research outputs, in light of local culture resistant to market-oriented reforms, and the lack of funding, that could result from the tightening of the fiscal policy.

Table 2.4: Improve Financing of S&T Sector

Implementing agency: Coordinated action: MSES, MELE, MFIN, -coordinated by NISC

Measures, instruments: selective financial instruments

Specific objectives:

- Increase public financial support for S&T and R&D both for public and private research performers
- Introduce selective fiscal and tax measures
- Implement a 3 per cent action plan based on Lisbon plan, with a structure of 1:2 between public and private investment
- Combine direct measures, fiscal incentives, guarantee schemes, risk capital
- Support development phases from seed to growth for knowledge-based companies
- Promote mechanisms where public S&T Institutions can exploit participation rights derived from IP invested in industrial products/services
- Financially support the growth of research groups to critical mass of EU significance and competitiveness, combine with mobility measures
- Introduce mechanisms for backflow of funds from successful supported projects
- Strengthen methodology for project evaluation within the system (BICRO, S&T projects)

Key outcomes: increase in volume of financing, increase efficiency of financing

Milestones: During 2007, finalize National 3 per cent Action Plan (by Committee for 3 per cent AP)

- WB financed STP is establishing Rudjer Innovations (RI) as technology transfer Agency fully owned by Rudjer Boskovic Institute.

Progress indicators:

T1. R&D expenditure as percentage of GDP, GERD

T2. Full-time equivalent R&D employees per 1000

T3. Number of patent applications per million population

T4. Number of CC/SCI publications per 1000 R&D employees

T5. Number of Science and Engineering (S&E) graduates per 1000 population

T6. Number of S&E PhDs per 1000 population

T7. R&D and innovation expenditure of companies as percentage of GDP, BERD

T15. Number and value of participation of Croatian researchers in international R&D projects

R&D in Croatia is carried out by public sector PROs, HEIs and on a limited basis in the private sector, mostly in large companies. Public sector approaches in the past have led to inefficient infrastructure and services that do not respond well to market needs. An emphasis on private sector participation therefore is important in the restructuring process, for the sector's ability to inject into the process efficiency, management capability, good governance, and cost-effective design and delivery.

The private sector is instrumental in making a shift from knowing to applying the knowledge for the benefit of the economy.

The following key aspects of SME competitiveness need to be addressed:

- ☆ Market limitations
- ☆ Profitability problems
- ☆ Supportive financial infrastructure
- ☆ Knowledge gap
- ☆ Innovations
- ☆ Clustering

Table 2.5: Intensify R&D in Industry

Implementing agency: Coordinated action: MSES, MELE, NISC, BICRO, RI, Croatian Chamber of Commerce

Measures, instruments: coordination through NISC

Specific objectives:

- ☆ Support for technology transfer into companies
- ☆ Increase number of companies doing R&D, increase intensity of R&D
- ☆ Intensify R&D demand from business sector in public RDIs and universities
- ☆ Attract FDI which consume local R&D
- ☆ Further develop matching grant schemes for development (SPREAD program within BICRO, etc.)
- ☆ Develop and promote new forms of cooperation between public research and industry
- ☆ Support mobility between business sector and public S&T network
- ☆ Set up programs that will stimulate private investments into S&T, where limited public funds would act as leverages and catalysts for private investment–support public–private partnership in S&T
- ☆ Interaction between public S&T and economy needs to evolve from R&D sponsorship to mutual participation in projects with common interest
- ☆ 'Scientists in industry' programs

Key outcomes: private sector doing more R&D for their development needs

Milestones: Croatian Chamber of Commerce to organize workshops for industrial companies

- ☆ Integrate successful R&D institutes in MSES network
- ☆ Tax benefits enabled by Law on Science and Higher Education

Progress indicators:

T9. Number and value of R&D contracts between SMEs or industry and universities or RTOs

T10. Number of S&T start-up SMEs

T11. Percentage of employees in high-tech services

T12. Percentage of employees in high-tech manufacturing and industries

T13. Sales of new-to-market or new-to-company products as percentage of total sales

T14. High-tech exports

The general public and specific stakeholders' influence should never be underestimated. Therefore select your target groups carefully: local industry, SMEs, institutions, public S&T sector, international industries, overall society and devise a promotion campaign aimed at raising awareness to the necessity of developing a culture of innovation.

Table 2.6: Public Acceptance – Promotion Campaign

Implementing agency: MSES, NISC, Professional Agency selected through public tender

Measures, instruments: Promotion campaign in various media, organize events for target groups, develop dedicated website

Specific objectives:

- ✰ Promote culture of innovation
- ✰ Generate awareness for commercial exploitation of academic projects
- ✰ Promote fact that standard of living and quality of public services can not be improved or even maintained in a non-competitive low-tech and low-profit economy
- ✰ Promotion of Croatian brands and names in S&T and R&D,
- ✰ Promotion of impact of buying domestic R&D on the development of society
- ✰ Promote R&D institutions and their services
- ✰ Provide dedicated website

Key outcomes: stimulated demand for R&D services

Milestones: Campaign to be launched September 2007.

Progress indicators:

T1. R&D expenditure as percentage of GDP, GERD

T5. Number of Science and Engineering (S&E) graduates per 1000 population

T6. Number of S&E PhDs per 1000 population

T7. R&D and innovation expenditure of companies as percentage of GDP, BERD

T8. Percentage of employees participating in lifelong professional education

Recognizing and Developing Priority Areas

For Croatia, the priorities that enable globalization of knowledge, scientific propulsion, economic efficiency based on the values of a humane society and those that directly support rapid progress of basic sectors of the economy, are the most promising. In 2005 strategic research priorities were proposed by an MSES expert group. Some of these priorities are of national interest and others are in line with EU Framework programs (FP6 and FP7).

The long-term priorities were:

- ✰ Knowledge-driven basic research
- ✰ Environmental protection and economic development of the Karst regions; Adriatic sea, coast and islands
- ✰ Agriculture; Biotechnology; Food
- ✰ Health

- ☆ Information and communication technologies
- ☆ Nanoscience; New materials, construction and new production processes
- ☆ Energy; Sources of alternative and renewable energy; Transport and security
- ☆ Social and human sciences; Croatian identity
- ☆ Social integration, learning and education; Lifelong learning

The short-term priorities were:

- ☆ Environment (Adriatic sea, coast and islands; Karst region)
- ☆ Health (Food; Agriculture; Biotechnology; Socials aspects of health; Health systems)
- ☆ Energy and Materials (Alternative and renewable energy; Bio-nanomaterials)
- ☆ Croatian Identity (Croatian contribution to culture, religion, art and sciences; Croatian language)

Within these priorities Croatia should indentify, encourage and develop highly specialized niches, through which it would become recognizable in the global knowledge society and competitive in the world market.

Conclusion

All around the world economic activity is shifting towards innovation and knowledge-driven industries. Knowledge and information have replaced energy and capital as the primary wealth-creating assets. The only competitive advantage any business, enterprise or an entire society can enjoy is its own process of innovation and the ability to derive value from knowledge and information. In this context, science makes vital contributions to knowledge-driven economies by generating new ideas and technology solutions. Equally important to the knowledge-based economy is the ability to turn scientific discoveries into successful commercial products.

Like other societies, Croatia could benefit tremendously from having a strong science and technology sector, therefore it is the Government's primary strategy to invest in science and technology development, to ensure that it is built on excellence and to maximize its contribution to long-term economic development.

The current S&T system practices and incentives structure do not sufficiently promote the commercialization of research outputs into technology development, practical applications or enterprise spin-offs, therefore, offering little benefit to Croatian society. The use of untapped knowledge resources could raise innovation and productivity in the economy, in turn smoothing the impact of the current financial crisis and offering new avenues for development in the 21st century.

References

MSES, October 2006, Science and Technology Policy of the Republic of Croatia 2006 - 2010, Zagreb, Croatia.

http://public.mzos.hr/Default.aspx?sec=2127

MSES, April 2007, Action Plan 2007-2010 for the Science and Technology Policy of the Republic of Croatia, Zagreb, Croatia.

http://public.mzos.hr/Default.aspx?sec=2127

Abbreviations

BICRO: Business Innovation Center of Croatia, a governmental technology funding agency

CAA: Croatian Accreditation Agency

CC: Current Contents

EU: European Union

GOVBOST: Government Body for Science and Technology, high-level sponsors for S&T Policy

HEI: Higher Education Institution

IPR: Intellectual Property Rights

MELE: Ministry of Economy, Labour and Entrepreneurship

MFIN: Ministry of Finance

MEPPPC: Ministry of Environmental Protection, Physical Planning and Construction

MSES: Ministry of Science, Education and Sports

MSTTD: Ministry of Sea, Tourism, Transport and Development

MSTQ: Measurements, Standards, Testing and Quality

NIS: National Innovation System

NISC: National Innovation System Council

OECD: Organization for Economic Co-operation and Development

PHARE: Pre-Accession funding program of the EU

PRO: Public Research Organization

RBI: Rudjer Boskovic Institute, the largest PRO in Croatia

RI: Rudjer Innovations, the technology transfer company of RBI

R&D: Research and Development

S&T: Science and Technology

SIPO: State Intellectual Property Office

STP: Science and Technology Project, a WB funded projected implemented by the MSES

WB: World Bank

Chapter 3

SEMAT: A Must for Science and Technology Policy Making

Seyed Omid Fatemi

Director

Iranian Research Institute for Information Science and Technology, Iran

E-mail: fatemi@irandoc.ac.ir

ABSTRACT

The need for science and technology (S&T) policy making has become increasingly important. The first and most important requirement of such capability is to have current, complete and correct science and technology information. The national S&T policy maker, the Supreme Council of Science, Research and Technology (SCSRT) of Iran, which is headed by the president and consists of main decision makers of the country in the research area, including several ministers, has ordered the development of a national current research information system, code named SEMAT, in 2006. The main reasons for such a system are to help top decision makers for the research process in the country and also to help researchers to access scientific information.

A group of representatives from SCRST is formed as the steering committee of SEMAT. For the first phase of the project, three sub-systems have been defined by the committee. In this paper the architecture of SEMAT and its main services are presented.

Introduction

The need for science and technology (S&T) policy making has become increasingly important. Today, the main asset for every country is the acquired science and technology in the country and to make the best out of this capacity there is a need to have active and current national S&T policies in place.

On the other hand, knowledge growth has been in exponential form in the last decade. It is reported that 80 percent of all science and technology findings are produced in the 20th century. And it is predicted that every five years they are doubled and in less than four years they become obsolete. Therefore, the need for a reliable, fast, efficient and convenient access to scientific information should be answered carefully.

These facts dictate the necessity of having research information systems which enable policy and decision makers to have all the required information and knowledge to make the best decision.

Imam Sadiq (peace be upon him) has said:

العَامِلُ عَلَى غَيْرِ بَصِيرَةٍ كَالسَّائِرِ عَلَى غَيْرِ الطَّرِيقِ لَا يَزِيدُهُ سُرْعَةُ السَّيْرِ إِلَّا بُعْداً الامام الصادق عليه السلام:

"The one who works without insight is like he who walks in the wrong way, speeding up would only make him farther (from his target)."

The supreme leader of Islamic Republic of Iran, Ayatullah Sayyed Ali Khamenei, in his lectures emphasizing the importance of research in the country has said:

"Definitely, research should be considered more....And the main requirement is to have a centralized database for information."

These requirements have been the main force behind systems such as Frida in Norway and NARCIS in Netherland. Looking at the goals and objectives of such systems, one could find many similarities in such systems. Grete Christina Lingjoerde introduces Frida as: "Frida is an integrated research environment for the documentation and presentation of research activities, research results and scientific competence. Data from Frida is used to generate statistics for research activities at Norwegian universities. Information provided by this system plays a major role in the annual funding of universities by the Norwegian Ministry of Education and Research." (Lingjoerde and Sjorgren, 2008).

For these reasons, the Supreme Council of Science, Research and Technology (SCSRT) of Iran which is headed by the president and consists of main decision makers of the country in research area including several ministers has ordered the development of such a national system in 2006. The main reasons of such system are declared as follows:

- ✰ To regulate the national scientific system
- ✰ To extract inter-organizational level of scientific information
- ✰ To develop on demand research system
- ✰ To develop international scientific collaborations
- ✰ To use scientific information and method for decision making.

SCSRT has declared the following objectives for implementing national current research information (SEMAT, in Persian):

1. National knowledge management;
2. To support developing national science and technology roadmap;

3. To help organizing the process of science and technology generation;
4. To support decision makers;
5. To define national research priorities;
6. To support and automate the process of assigning public budget for research institutes;
7. Evaluating and ranking research institutes based on defined criteria;
8. Increasing collaboration level among researchers throughout the country;
9. To provide fast, easy and reliable access to scientific information for researchers;
10. To improve the quality of research;
11. To prevent unwanted repeating research works;
12. To promote the standards of generating, organizing, processing, disseminating and retrieving scientific information;
13. To provide an area for industry to announce their needs and for researchers to propose their solutions.

In order to start implementing the system a group of representatives from SCRST is formed as the Steering Committee of SEMAT. During the last three years, this group is managing the SEMAT project. They have developed the RFP for the first phase. A private company has been selected for developing the system. The first phase of SEMAT includes three main sub-systems:

1. The sub-system of collecting the information;
2. The sub-system for accessing information based on access levels;
3. The sub-system for generating managerial and informative reports.

In this paper, the conceptual model and the architecture of the system is presented. In the next section, for the inception phase, the current situation of research processes is presented. The architecture of SEMAT is presented in the next section followed by conclusions.

Inception Phase

In order to start the inception phase, at the first step, the research institutes are defined and introduced. Then the pilot institutes for SEMAT project are selected and the current research processes in the pilots are presented.

Research Institutes

The main research body in Iran are formed by universities and research institutes either Governmental or Non-Governmental. In the Governmental part, the two main ministries in charge for research are Ministry of Science, Research and Technology (MSRT) and Ministry of Health and Medical Education (MOH). The largest non-governmental university is Islamic Azad University (IAU). The universities and research institutes are independent bodies and have their own management of research process. The policies and standards are set by MOH for health section, MSRT for other universities and the management body of IAU for all IAU branches.

Selected Pilots

Since during the deployment of SEMAT project, scientific information is required to be communicated among research institutes and SEMAT, the Steering Committee has decided to select pilot institutes for the deployment phase of the project.

Current Status of Research Processes in Institutes

We have selected Enterprise Architecture Model FEAF (Federal Enterprise Architecture Framework) shown in Figure 3.1 to evaluate and rank institutes and to perform the inception phase. The model is built on common business practices and designs that cross organizational boundaries and is suitable for our purpose. The FEAF provides an enduring standard for developing and documenting architecture descriptions of high-priority areas (Thomas, 2001).

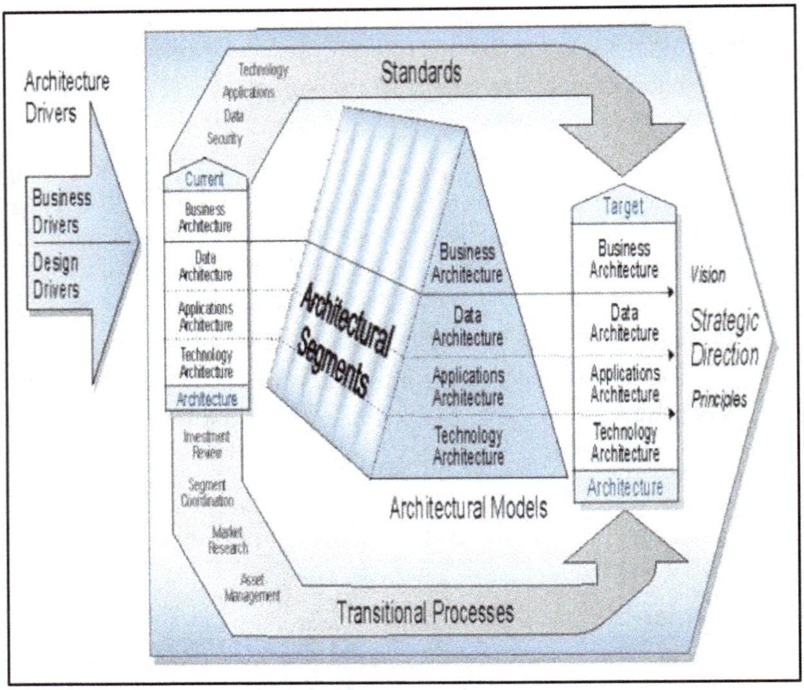

Figure 3.1: Structure of the FEAF Framework (Rob C. Thomas 2001)

To be able to define the best suitable model for SEMAT, the current processes of research institutes have been identified and modeled. Then four layers of information have been presented for every pilot which is as follows:

- ☆ Business layer
- ☆ Data layer
- ☆ Application layer
- ☆ Technology layer

SEMAT Architecture

Having done the inception phase, we analyzed the result and updated the main requirements as follows:

1. Integrating all national research meta-data;
2. Collecting research record sets;
3. Registering scientific and technological information;
4. Providing collaboration environment;
5. Searching through information.

Based on the heterogeneity and the complexity of the system, service oriented architecture is selected and all the communications throughout the whole system are performed through receiving and transmitting messages in an Enterprise Service Bus.

We have performed the following steps in order to implement the system:

1. Research meta-data entities have been identified and the message format and the lifecycle of their corresponding messages have been defined.
2. Meta-data entities are being generated at the possible earliest time in the research institutes.
3. All messages are sent from the institutes to the National SEMAT.
4. SEMAT receives the messages and processes them.
5. SEMAT manages the messages and stores the validated and processed meta-data.

Based on this architecture, the institutes are categorized in three groups:

1. Those that don't have any research information system. In this case, SEMAT provides an environment for the institutes to transmit their messages.
2. Those that have a research information system, but it is not and cannot become compatible with SEMAT. In this case, messages are generated asynchronously by these systems and are imported to SEMAT.
3. Those that have a compatible research information system. In this case, systems communicate with SEMAT and messages are going back and forth among these systems.

Research Information Integration

The best method to add the ability of integration of collected data from all research organizations is to define a common data model. Our proposed data model is based on our newly introduced Research Information Format called IRIF, Iranian Research Information Format. To design an acceptable IRIF, three steps have been taken:

1. Having all existing data models of the pilot institutes, we have first extracted all common data fields among them.
2. In the second step, the above data model is compared with CERIF (Common European Research Information Format).

3. As the final step, the steering Committee decided on the differences found in the second step and selected the final data model as the final Iranian Research Information Format (IRIF).

The Proposed Method

As discussed earlier, the ideal model of the national SEMAT is to be able to communicate directly with institutional SEMATs. In this case, during the generation of research information in institutes, the required data (usually meta-data) would be sent to the national SEMAT. This method will ensure the completeness, being up-to-date and validness of registered information. In our proposed method all communications are handled through passing messages in an Enterprise Service Bus. The messages are generated in each institute by a message client and in necessary intervals the message client sends the data to the server. The block diagram of the architecture is shown in Figure 3.2.

SEMAT Services

Having designed the architecture, we have defined different services for SEMAT as shown in Figure 3.2.

These services include:

1. Query and Search service: This service is to perform queries in SEMAT. Full-text search, thesaurus aid in search and structural queries are among the features of this system.
2. Budget management in research institute: In order to perform national budgeting and public resource allocation for research institutes, this service is defined.
3. Coding services: To establish the registration system for research entities and to facilitate integration, a coding service is designed. This service assigns a unique code for every instance of research entities and this code is used as the identifier in the whole system.
4. Validation services: This service is in charge to evaluate, modify and validate data before being stored in SEMAT repositories. Both accuracy and consistency controls are done in this service.
5. Portal and community services: The single point of connection to SEMAT for the users to use the services is through this service. This service enables users based on their access rights to access information and make use of the services. Every user in the system has his/her own space in the system, and is able to personalize the environment. General services such as email, forums and RSS feeds are also provided.

Conclusions

Based on the necessity of managing all research information in Iran and also the lack of generally agreed standard for this data the project of National SEMAT has been initialized by SCRST. In the first phase of the project, the main effort has been

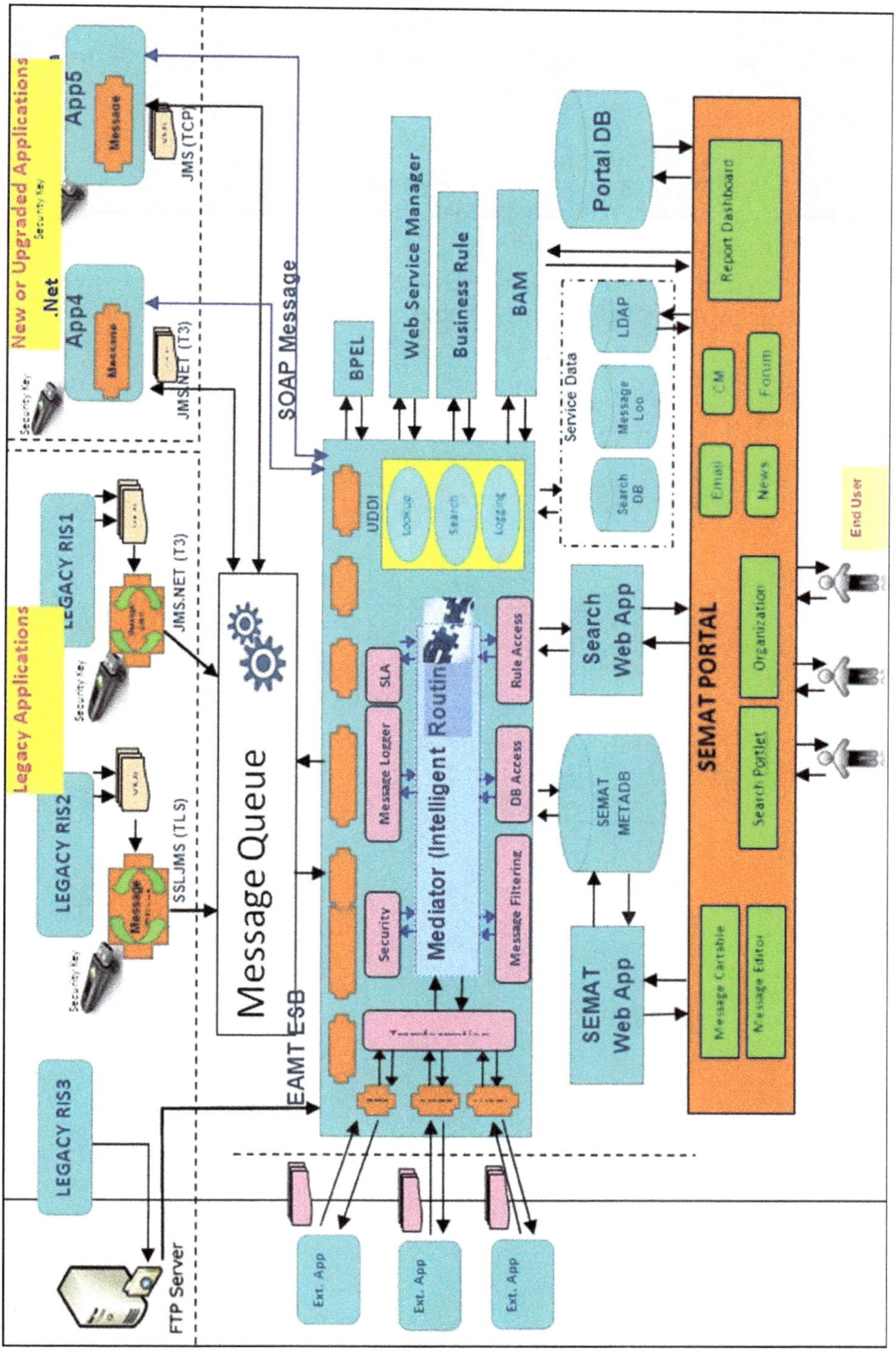

Figure 3.2: SEMAT Architecture

put for collecting all information using a dynamic method which connects the institutes to SEMAT. The system is based on message passing and each message has its own life cycle. The first implementation of the system has shown the effectiveness of the system in using and managing research data.

References

Grete Christina Lingjoerde and Andora Sjørgren, 2008. Quality assurance in the research documentation system Frida. *Cris 2008. 9th International Conference on Current Research Information Systems.*

Thomas, Rob C., 2001. *A Practical Guide to Federal Enterprise Architecture.* http://www.gao.gov/bestpractices/bpeaguide.pdf

Chapter 4

Using Stakeholder Engagement to Bridge the Gap between Science, Technology and Innovation (STI) Policies in Implementing the Kenya Vision 2030 for National Development

Frank N. Sawanga Ndakala

Directorate of Research Management and Development,
Ministry of Higher Education, Science and Technology, Kenya
E-mail: ndakala@scienceandtechnology.go.ke, ndaxsawanga@yahoo.com, ndakalasa@yahoo.com

ABSTRACT

Kenya abandoned her long standing "Poverty Reduction Strategy Papers (PRSPs)" in 2003 when a new government took power. Throughout the 90's Kenya's economy plummeted because of the limited consultation done during PRSPs development and the negative perception by the public. To develop a positive perception the new government widely engaged stakeholders in developing a new economic blue print "The Economic Recovery Strategy for Wealth and Employment Creation 2003-2007." In 2008 this was replaced by Kenya Vision 2030 which aims to transform Kenya into a newly industrialized middle – income country by the year 2030. As a product of the government and stakeholders, the vision has three pillars, economic, social and political pillars. Science, Technology and Innovation (ST&I) forms a fundamental element for ensuring the success of Vision 2030, hence ST&I sector needs a new stakeholder driven impetus. The recent promulgation of the new constitution as projected in the political pillar of the vision is a clear manifestation of benefits of consensus building between stakeholders and the government. Such major paradigm shift has enhanced the participation of various interested groups who previously had no opportunity of influencing government decisions but can now express their opinion through various lobby groups in a free and fair environment without interference.

This paper attempts to examine the mechanisms that the government has established to enhance stakeholder participation in policy formulation, development and implementation. The main focus will be on the experiences and lessons learnt in developing the ST&I policies like the ST&I Bill, and the Bio-safety Bill. It concludes by highlighting key lessons from Kenya that can help other African countries to foster increased stakeholders participation in decision making for policy and public debates in matters related to Science, Technology and Innovation.

Keywords: *Vision 2030, ST&I, Stakeholders, Bio-safety, Policy, Research Institutes.*

Introduction

The interest-group theory of government reveals that public policy emerges from the interplay of various interest groups in any given country (Stigler, 1971; Peltzman, 1976). Well-organized and moneyed interest groups are able to influence policy choices in their favour. Weak and poorly organized groups are generally ignored by policy makers, as they do not have the necessary leverage to compel policy makers to take them seriously. It therefore follows that, interest groups compete with each other in seeking transfers or preferential treatment from the government. For example, large scale farmers will lobby the government to put in place policies that lower their cost of production (subsidize inputs) and raise the price of their products (guaranteed minimum prices). In such cases, the general taxpayers and consumers are the losers from such policies, while the farmers receive the gains. It is unlikely that small-scale farmers will be able to influence such policies. In most cases, it will be the large, well-organized farmers that are capable of influencing government policy. Likewise, well-organized, large manufacturers may be able to demand favourable policies that benefit them, but small-scale operators (*e.g.* 'jua kali') find it difficult to influence in their favour.

In Africa, the most powerful interest groups are primarily tribal or ethnic based. They compete for government favours in order to improve the well-being of their tribe at the expense of other tribes. A government that responds to such ethnic interest is unlikely to adopt policies that support growth. Historical evidence shows that economic growth has been sustained only in countries where there is an all inclusive interest group active participation in policy formulation. It is because of the power play involving stakeholders with various levels of interest and power that the Kenya government came up with structures that can enable interest groups participation in policy development. To enable this participation, the structures in place can facilitate the sharing of ideas, information and provide feedback and resources. These structures are referred to as Public Sector Stakeholder Partnerships (PSSPs). The policy document for Public Sector Stakeholder Partnerships (PSSPs) was therefore developed in support of this initiative (PSSP, 2007).

The government of Kenya has developed Vision 2030 whose main objective is to turn Kenya into a globally competitive and prosperous Nation with a high quality life by the year 2030. Kenya Vision 2030 is anchored on three key pillars: Economic, Social and Political. Each pillar has a number of strategies designed to deliver different components of the Vision. Science, Technology and Innovation (ST&I) is one of the fundamental elements for ensuring that the goals in the three pillars of the Vision 2030 are realized (Government of Kenya, 2008). Lessons learnt from the previous economic blue prints of the 1990s; *i.e.* the Poverty Reduction Strategy Papers (PRSPs) and the Economic Recovery Strategy for Employment and Wealth Creation (ERSEWC, 2004), pointed out to the increasing need for stakeholders involvement in developing sustainable policies. Therefore, the development of PSSP in 2007 aimed at bringing to fore a mechanism for enabling stakeholder participation in developing policies that support the attainment of the three pillars of Vision 2030. This paper attempts to look how the implementation of PSSP has helped in developing ST&I policies to support the realization of Vision 2030. The most recent policies to be discussed include the ST&I policy and the Bio-safety Bill. The challenges encountered during the development of these policies will be discussed and the way forward proffered.

Historical Background of Stakeholder Engagement in Kenya's Policy Development

The Government of Kenya subscribed to the Poverty Reduction and Growth Facility (PRGF) in 2000 and embarked on the preparation of the Poverty Reduction Strategy Paper (PRSP) at the same time. This preparation was undertaken through wide-ranging consultations and dialogue in order to build consensus on priority actions and activities necessary for economic growth and poverty reduction. (ERSEWC, 2004). The PRSP was preceded by the Interim Poverty Reduction Strategy Paper (IPRSP) released in 2001. The IPRSP only involved limited consultations at the national level. The PRSP consultations followed a three-tier approach: national, provincial and district levels. The stakeholders in the consultations included the Private Sector, Civil Society, the Development Partners and local communities. A national steering committee that included all the stakeholders was formed to spearhead the process and ensure inclusion at all levels. The consultations went as far down to the divisions, locations and villages in some of the districts. Participatory Poverty Assessments (PPA) were carried out in ten (10) sampled districts. The District PRSP reports and PPA reports together with inputs from the Sector Working Groups were synthesized into the PRSP (2001-2004). The PRSP formed the basis for the 2002/03 budget.

In December 2002, a new government under the National Rainbow Coalition (NARC) took office and prepared Economic Recovery Strategy (ERS), focusing on reviving the economy and creating employment. The ERS presents a road map for economic recovery during the period 2003-2007. The ERS takes into account existing Government policy documents, particularly the PRSP and NARC's Manifesto and Post-Election Action Plan (ERSEWC, 2004). The development of ERS was also a result of wide-ranging consultations with stakeholders. The stakeholders included: parliamentarians; trade unions; professionals; financial institutions; industrialists;

ASALs; development partners; civil society and government. The strategy therefore embodies the views and aspirations of all Kenyans.

To enhance participation of stakeholders in future national policies, the Government had to put in place certain structures. These include the establishment of The National Economic and Social Council (NESC), Ministerial Stakeholder Forums (MSFs) and Task Forces (Government of Kenya, 2007). NESC advices the government on polices that are required to address socio-economic issues at all level of the citizenry. It is a standing committee comprising of eminent persons with diverse experience, knowledge and skills. The president chairs the council. It provides a forum in which the government, private sector, civil society and the labour unions can discuss and identify policy issues and recommend to the Government, implementation of such policies. MSFs have been formed in several ministries. They comprise of key stakeholders representatives and their main task is to participate in policy formulation, research, advocacy, resource mobilization, project implementation and monitoring and evaluation. Task forces: From time to time the government appoints task forces to consider specific issues such as education, land policy etc. These task forces are usually comprised of recognized authorities in specific field from both the public and private sectors.

Current Trends in Public Participation

Citizen input into science and technology decision making currently takes many forms: community-based and participatory research, consensus conferences, and scenario workshops.

Community-Based Research (CBR)

Community-based research (CBR) is a collaborative partnership between researcher and community. CBR is conducted by, or in participation with, the community that is affected by the problem in question. Practitioners who challenged conventional top-down approaches to policy development pioneered CBR several decades ago. In the early 1970s, researchers– primarily in Asia and Latin America– began to question the reductionism of most research, and its inability to solve the myriad societal problems. (Chopyak and Levesque, 2002). Working with marginalized communities, researchers began to collaborate with community members in designing and implementing research projects that had direct relevance to their struggles. The idea of CBR grew during the 1980s as those involved in international development grew increasingly frustrated with their inability to solve problems related to community development, education, health and poverty. Development practitioners began to work closely with researchers and community members using participatory research methods as a way of developing effective solutions to many of the problems they face.

Because all communities have more problems and fewer resources to solve them, there has been a rise of public, non-governmental participation in research, science and technology decision making. Involving expertise or experience from universities or other sectors is a method of leveraging intellectual and financial resources perhaps not otherwise available. There are increasing numbers of success stories of how leveraging resources, knowledge, capacity and history have resulted in mutually beneficial results for the majority of stakeholders in a project (Sclove *et al.*, 1998).

Consensus Conference

The overemphasis of the need for the public support for scientific endeavours has been termed as the 'deficit model' or hierarchical approach to science communication. In this context it is assumed that the general public will simply accept new technologies provided that the science is explained to them and non-acceptance of such technology is often considered as an irrational response, rather than an expression of legitimate concerns" (Outram, 2010). Recent survey has established that this model focuses on measuring the public's scientific literacy such as knowledge of S &T facts and understanding of S&T processes (Reddy *et al.*, 2009a). It has been argued, and correctly so, that the surveys being used in this model "are structured in a way that will suggest that the public is deficient in knowledge and positive attitudes" (Reddy *et al.*, 2009a). Consequently, the public ends up being excluded from participating in S&T matters. To the contrary, there is evidence on how creating a dialog with citizens about science and technology issues has resulted in better decisions.

One good example of engaging the public to develop better policies is through the consensus building conferences. The consensus conference process allows citizens to make policy recommendations about specific scientific or technological developments, and those recommendations are then included in the process by which the legislature makes its decisions.

In Denmark the consensus conference process brings together 12–15 citizens to learn and deliberate about a complex science and technology issue (Chopyak and Levesque, 2002). The lay panel then hears from and cross-examines a panel of experts. This information is used by the lay panel which develops a policy statement on the issue, which is presented to policy makers and the general public (Einsiedel *et al.*, 2001). The impact of the consensus conference process on policy decisions differs depending on the social and political context in which it takes place. In the Danish example, the process is supported and coordinated by the Danish Board of Technology, which reports directly to the Danish Parliament. In this context, the recommendations developed by the lay panel are directly fed into the decision-making process. It is clear that the consensus conference process has consistently demonstrated that when citizens are given relevant information, they can make coherent and well-thought-out recommendations on complex science and technology issues.

Scenario Workshop

Developed in the early 1990s, the scenario workshop process is a participatory urban planning methodology that brings together key stakeholders to discuss and evaluate the impact of multiple technological choices on a particular setting. The process is based on four scenarios that describe the day in the life of a resident, 20 years from now. Each scenario describes alternative ways of solving particular problems (such as energy, water, housing and transportation). The scenarios are designed as visions, not predictions. Stakeholders then come together in a workshop to develop their own vision of the future, using 'role play' and the four scenarios as the basis for discussion. The goal of the process is for participants to agree upon a vision and articulate a set of action plans to achieve it (Chopyak and Levesque, 2002). The scenario workshop process requires significant preparation and planning. Recruiting a diverse group of stakeholders to participate is key to the success of the process. Its potential is to bring together local government, grassroots activists, science and technology experts, and citizens to create a vision for their community. It allows all groups to see different point of view(s) and to deliberate about the technological choices a particular locale is making. Because of its complexity, the scenario workshop process is not as widespread as the consensus conference, and its adoption has been predominately within European settings.

The Policy Making Framework in Kenya

In Kenya, just like other developing countries, policies are determined through the general framework of national economic planning. The Ministry of Planning and National Development (MPND), through their periodic sessional papers on National Development Plans, outline the country's priority areas and activities for economic development and undertake to monitor project implementation by the various responsible Ministries. To deliver its mandate, MPND works with the Kenya Institute for Public Policy Research and Analysis (KIPPRA) in the production of sessional papers and national policies. KIPPRA is an autonomous institute under MPND. Its primary mission is to conduct public policy research, leading to policy advice. KIPPRA's mission is to produce consistently high quality analysis of key issues of public policy and to contribute to the achievement of national long-term development objectives by positively influencing the decision-making process. These goals are met through effective dissemination of recommendations resulting from analysis and by training policy analysts in the public sector. KIPPRA therefore produces a body of well-researched and documented information on public policy, and in the process it assists in formulating long-term strategic perspectives. KIPPRA serves as a centralized source from which the government and the private sector may obtain information and advice on public policy issues.

In 2008 the Ministry of Planning and National Development (MPND) spearheaded the development of Kenya Vision 2030, which the country's economic blue print. The Kenya Vision 2030 is to be implemented in successive five year Medium Term Plans, with the first such plan covering the period 2008-2012. Different projects and programmes (flagship projects) have been scheduled to be implemented by different ministries within different phases of 5 year periods. (Government of Kenya, 2007). All Ministries are expected to concentrate on the prioritized flagship projects. Proper implementation of these projects demand that ministries streamline their policies and even initiate the development of new policies as part of attaining Vision 2030.

Considering that Kenya's economic reform effort had previously been hampered by the lack of adequate consultation between the government and the public. This meant that the Ministry of State for Public Service was to come up with the Public

Service Reform Strategy (PSRS). The main objective of this strategy was a public service that is re-energized to facilitate the delivery of Vision 2030 goals. There was need for the public service to transform itself and start working with stakeholders in policy development and implementation. The Ministry of State for Public Service therefore initiated the development of a policy on engaging stakeholders in the policy issues, hence Public Sector Stakeholder Partnerships Policy (PSSP). The PSSP provides a framework for the inclusion of every Kenyan's input in policy making and provide opportunities for stakeholders to be involved in public service delivery (Government of Kenya, 2007).

Since Science Technology and Innovation (ST&I) forms a key fundamental element for the attainment of Vision 2030, the Ministry of Higher Education Science and Technology (MOHEST), has the mandate to initiate policies that can contribute to the realization of Vision 2030. The policy development mechanisms in MOHEST are hinged on the Science and Technology Act of 1989. The Act mandates the National Council for Science and Technology (NCST) to advise the MPND on all scientific documentation, statistics, surveys and general information, with the specific objective of establishing the necessary machinery for the collection and dissemination of scientific and technological information (Republic of Kenya, 1993). For a long time since independence, despite the evident potential of investing in R&D activities, Kenya lacked a strategic plan for a coordinated R&D initiative. To address this issue, the government through the 1970 National Development Plan, considered three main options for the formulation of scientific and technological policy by the Ministry of Economic Planning and Development and the office of the President: Administrative creation of a bureau of statistics within the Ministry of Finance and Planning; creation of a Ministry of Science and Technology; establishment of a statutory body for technology policy-making and coordination of research. The 1974-1978 Development Plan proposed to establish a National Council for Science and Technology (NCST) as a statutory body with the following objectives:

- determine priorities for scientific and technological activities;
- advise the government on a national science policy, including general planning and the assessment of requisite financial resources;
- ensure the application of scientific research activities to the development of agriculture, industry and social welfare;
- ensure corporation and coordination between various agencies involved in making the national science policy;
- promote public confidence in scientific expenditure and create an atmosphere conducive to scientific activities.

Actors in Science and Technology Policy Making in Kenya

Most of the science and technology activities in Kenya have been in the applied science disciplines like electronics, engineering, medicine and public health. However, most government policies have stimulated the development of S&T in the fields of agriculture and medicine at the expense of other disciplines. There are six strategic national research institutes mainly in the fields of agriculture and medicine. Even with the recent explosion of life changing technologies in the fields of engineering and electronics there is still no national research institute to support the sector. This scenario has therefore, greatly determined the main actors in the S&T policy-making structure. The six national research institutes that are always in the forefront of influencing ST&I policies are the Kenya Medical Research Institute (KEMRI), Kenya Agricultural Research Institute (KARI), Kenya Industrial Research Development Institute (KIRDI), Kenya Forest Research Institute (KEFRI), Kenya Marine and Fisheries Research Institute (KEMFRI) and Primate Research Institute (PRI). Universities which collaborate with the research institutes do on their part offer academic programmes that support the growth of ST&I in these two major sectors. In addition, the government has facilitated the growth of S&T in the agricultural sector by creating commodity research foundations namely; the Kenya Sugar Research Foundation, Coffee Research Foundation and the Tea Research Foundation. The prevailing government policy that favours the growth of S&T mostly in agriculture and medicine has been replicated by the development partners and non-governmental organization involved in research.

In agriculture, the main stakeholders include: the Ministry of Agriculture, various public agricultural research institutions such as KARI, the National Plant Breeding Station at Njoro, the Kenya Plant Health Inspectorate Services (KEPHIS), the Veterinary Research Laboratories, the Kenya Tea Development Authority (KTDA), the Coffee Board of Kenya, the Pyrethrums Board of Kenya and the Kenya Sugar Board. In human and livestock health, the Kenya Medical Research Institute (KEMRI), the National Public Health Laboratory, the Division of Vector Borne Diseases (DBVD), the Pharmacists and Poisons Board, the National Blood Bank, the National AIDS Control Council (NACC), the Division of Malaria Control and the Kenya Medical Association are the key players. In the wildlife management, key players include the Kenya Wildlife Services, the Kenya National Museum and the Institute of Primate Research. In the industrial applied sciences, the Kenya Industrial Research Institute (KIRDI), the Kenya Bureau of Standards (KEBS) and the Kenya Intellectual Property Institute (KIPI) form the key players.

Indigenous training and R&D establishments for scientists and technicians (and technologists) at the higher education level and tertiary institutions include both public and private universities, national polytechnics, technical education training institutes and Kenya medical training institute. Besides the national institutions, international research organizations sited in Kenya such as the International Centre for Insect Physiology and Ecology (ICIPE), the International Council for Research in Agro-forestry (ICRAF), the International Livestock Research Institute (ILRI) and World Heath Organization (WHO) are strategic stakeholders in Kenya's S&T stakeholders. International Development partners like Centre for Disease Control (CDC), Walter Reed Institute, Japanese International Cooperation Agency (JICA), United States Assistance for International Development (USAID) participate in various aspects of R&D activities. The African Center for Technology Studies (ACTS), an Inter-governmental organization, African Capacity Building Foundation (ACBF), African Medical Research Foundation (AMREF), the Consortium for National Health Research (CNHR), African Academy of Sciences(AAS) play significant roles in policy research on matters pertaining to science and technology in Africa.

The Ministry of Higher Education, Science and Technology (MOHEST) is responsible for government policy on S&T. It works in liaison with the NCST, which has mandate to ensure the cooperation and coordination between various agencies involved in making the national science policy, among other objectives.

The National Council for Science and Technology (NCST)

The National Council for Science and Technology (NSCT) is a semi autonomous agency under the Ministry of Higher Education, Science and Technology (MOHEST). It was established in 1977 by the S&T Act Cap 250 Laws of Kenya with the following mandate: to determine priorities for scientific and technological activities in Kenya in relation to the economic and social policies of the government and its international commitments. The core functions of the council are: advise the government on national science policy, including general planning and the assessment of the requisite financial resources; advise the government on the scientific and technological requirements for the conservation of the natural and social environment in Kenya; consider and advise generally on all scientific activities, advise on suitable organizational arrangements for planning, managing and coordinating scientific activities at various level, including the setting up of new research establishments and technical service; among other science and technology advisory functions. NCST, therefore offer scientists, researchers and the Government of Kenya, a forum where matters related to science and technology for development can be discussed and recommendations can be made to the Government for implementation. In order to achieve the mandate, the council is expected to develop policies, which can guide different sectors on scientific matters. However, this mandate is rather wide and requires some streamlining. First, it is not very clear if the financial capacity of the NCST (obtained through government budgetary allocation) is adequate for the practical execution of the tasks bestowed upon it. Secondly, the NCST needs personnel composition covering all the relevant disciplines of S&T fields. It is only through the availability of such capacity that it will be able to perform the necessary tasks.

Science and Technology Policy Making: Lessons from Kenya

The Science, Technology and Innovation Policy making process can best be illustrated by looking at the way the ST&I Bill 2009 and Bio-safety Bill 2009 were developed.

The ST&I Bill, 2009

The ST&I policy was drafted the way back in 2005. Although stakeholders were involved, the policy was not widely discussed by the public. In June 2009, the second National Conference for validation of the ST&I policy was organized and the policy document presented to the Cabinet before June 2009. The Attorney General of Kenya prepared the ST&I bill for cabinet approval, which was then presented to National Conference.

The ST&I Act repeals the S&T Act 1977 and creates: National Commission for Science, Technology and Innovation (NCST&I); Kenya National Research Fund (KENREF) and Kenya National Innovation Agency (KENIA). Research, Technology and Innovation MTEF Budgetary Sector was created in 2008/09 F.Y. In 2010, the ST&I was signed into a law by the president of Kenya.

The Bio-safety Bill 2009

The Bio-safety Act 2009 was approved by the President of the Republic of Kenya earlier in 2009. The Act lays down legal and institutional frameworks for governing modern biotechnology in the country. The Bio-safety Act 2009 created the National Bio-safety Authority (NBA) for overall supervision and control of the development, transfer, handling and use of genetically modified organisms for research or commercial purposes. During that year, the NCST facilitated the establishment of NBA secretariat. Unlike the ST&I Bill, the Bio-safety Bill generated a lot of public debate, which are evident in the processes to be discussed below. The challenges that were encountered in developing the Bio-safety Policy and latter on, the Bio-safety Bill will be discussed too.

Developing the Bio-Safety Bill: Kenya's Experience

The activities leading to the enactment of the Bio-safety Act 2009 benefited from concerted efforts of various task forces under the auspices of NCST. The NCST initiated this process in the early 1990s, with the support from the Government of Netherlands. In 1993, the task force sent a delegation to Harare to participate in a regional workshop on Bio-safety, as a part of the capacity building initiative (ABSF, 2003; Thitai, 2000).

To begin with, the NCST led the review and consolidation of all existing legislation related to biotechnology into a report (Thitai, 2001). The report was subjected to a thorough review by both local and international experts. The edited draft, dubbed "The Regulation and Guidelines for Bio-safety and Biotechnology in Kenya", was finally unveiled by NCST in 1998. This paved the way for establishment of the National Bio-safety Committee (NBC) and provided guidelines for the creation of Institutional Bio-safety Committees (IBC) in institutions conducting biotechnology research and development. The process received major financial and expertise boost from the Convection on Biological Diversity (CBD) fund under the UNEP-Global Environment Facility (UNEP/GEF), World Bank and the Netherlands (Thitai, 2000). These institutions facilitated most of the capacity building activities, which created a pool of local experts hence providing a major impetus to the Bill drafting process. These experts later took charge of the drafting and reviewing of the National Bio-safety Framework (NBF) in 1999. The drafting process got tedious, rigorous and nerve-racking as key interest groups got involved. There was a growing need to overcome barriers between scientists and non scientists though consensus building among stakeholders.

In 2002, the African Biotechnology Stakeholders Forum (ABSF) was formed to spearhead public awareness on modern biotechnology and bio-safety issues. The ABSF, which was hosted at KARI, became the hub of biotechnology awareness creation and knowledge sharing in the country by hosting the Kenya Biotechnology Information Centre (KBIC). KBIC was later christened the Eastern and Central Africa Biotechnology Information Centre (ECABIC) to reflect its regional mandate. It formed part of a global network of biotechnology information centres (BICs) ran by ISAAA Global Knowledge Centre (KC) on crop biotechnology based in the Philippines.

The second phase of the UNEP/GEF programme saw a strong push to have a comprehensive Bio-safety Law by like minded institutions, which had formed themselves into a close–knit network of biotechnology stakeholders under the ABST. To that end, a series of consultative workshops were held to draft the biotechnology policy, the Bio-safety Bill, national biotechnology strategy and to develop procedures for handling GMO applications.

The year 2001 marked the start of the actual drafting process, a highly interactive affair that involved a wide range of stakeholders. This was a transparent and participatory process, judging by the number and diversity of institutions and individuals who were involved. They included government representatives, parliament, universities, civil society organizations, media, industry, religious groups, farmer associations, development partners, UN agencies, research institutes, regulatory authorities and regional bodies such as ASARECA, BioEARN and the East African Community (ABSF, 2003).

A range of issues that needed to be addressed before the drafting was done, were identified during the stakeholders' workshops. Most importantly, there was a big gap between the scientists and non scientists due to the technical language that only few could understand. That created an impression that scientists did not want to share their knowledge with the public. A section of the stakeholders were very much concerned about the raging polarized debate on the merits and demerits of modern biotechnology, especially emanating from Europe. One group was opposed to the Bill because of the fear of introducing foreign GMOs into the country. Another controversy was on whether it was possible to enact a Bill without the policy. There was also contention over which government department should be made responsible for regulating biotechnology.

The team of experts drafted the Bill in July 2002, which was widely circulated among key stakeholders for comments and inputs. However, the Bill could not progress any further because Parliament was prorogued to allow for election latter in the year. There was a formal meeting in March 2003, where a detailed review of the draft was conducted to produce a fine tuned version. In April 2003, a week long stakeholders' meeting was convened to discuss this refined draft. By August 2003, the draft was ready to be presented to Parliament upon approval by the Cabinet. However the newly formed National Alliance Rainbow Coalition (NARC), Government and new parliament had to take time to form committees. In addition the new members of parliament needed to be sensitized on Bio-safety issues before they could discuss anything on the same. Now, new parliamentarians, the only easy thing to do was to stir debates on non technical issues. Therefore, politics dominated in most public debates throughout 2004 and the biotechnology scene was not spared. That year was marked by the clamor for a new constitution. The resultant tension seriously affected Government operations and the Bill was in limbo for a full year.

The NCST teamed up with ABSF, ISAAA *Afri*Centre through the Kenya Biotechnology Information Centre and KARI, to reach out to various key target audiences. For example, during 2004, a study tour was organized for three parliamentary committees; Health, Agriculture and Education, Science and

Technology. The experts interacted closely with the legislators and a lot of information was shared during those interactions, while views of the stakeholders were collected and collated. These views greatly enriched the draft Bill. Information materials were also produced and widely disseminated through the mass media, scientific exhibitions, farmer field schools, conferences, workshops electronically. These efforts were later complemented by other institutions such as the African Agriculture Technology Foundation (AATF), African Harvest, public universities and the Ministry of Agriculture through a new outfit, the Bio-safety Consortium, which consolidated the efforts of all the groups that had a stake in the Bill.

In 2005, after the constitutional referendum, there was fallout in Government and this resulted in the formation of new Ministries including the Ministry of Science and Technology. The new institutional arrangement also affected the process. The whole of 2005 was spent reaching out to different stakeholders and collecting their views and inputs into the already drafted National Biotechnology Development Policy and Bio-safety Bill. After several consultations, the minister for Science and Technology finally tabled the refined drafts to the Cabinet and they were approved in September 2006. There was much excitement among key stakeholders save for a few dissenting voices from a consortium, of civil society groups under the auspices of an outfit calling itself the Kenya GMO Concern (KEGCO).

The approval of the Bio-safety Bill paved the way for the Ministry of Science and Technology to forward it to the Attorney General for publishing in the Kenya Gazette. This would also give the public 21 days to respond to it as stipulated in the constitution. This however, did not take place immediately, due to a huge backlog of other Bills that were awaiting publication. On the other hand, KEGCO held meetings and used the media to discredit the Bill, arguing that it was hurriedly drafted and inadequate to protect against multinationals, who they claimed wanted to dump GMOs in the country. In August 2007, the group held huge demonstrations in Nairobi against both the policy and the Bill, with participants drawn from Uganda, Rwanda, Zambia, Ethiopia and Madagascar. They urged the government to postpone the debate on the Bill until after the impending 2007 general election. However the Attorneys General's office finally published the Bill in the Kenya Gazette, in 2007. Parliament passed the bill without any hitch. In January 2008, there was wide spread violence in Kenya resulting from the disputed presidential election of December 2007. When the violence subsided and the reform agenda put in place by the prominent African personalities, there was a flurry of capacity building and awareness creation activities targeting new members of parliament. On September 2008, about 10 MPs, parliamentary staff and representatives from ISAAA AfriCentre, NCST, ABSF, Programmes for Bio-safety System (PBS) and the Seed Trade Association of Kenya (STAK) held consultative meetings to discuss the way forward with regard to enactment of the Bill in view of the rekindled polarized public debate. This meeting also made aware that nearly 80 per cent of MPs who were sensitized in the previous parliament had lost their seats. Therefore, additional sensitizations were to be organized for members of parliament.

During the new sensitization workshops, the MPs were briefed about the lapsing of the Bio-safety Bill 2007. The Bio-safety consortium made a strong case in support of the Bill and informed the legislatures that the enactment of the Bio-safety Bill 2008 into laws was crucial in facilitating acquisition of relevant technologies to address the food insecurity and climate changes that the country faces. Of particular interest to MPs was a policy brief that had summarized the Bill and identified its relevance towards the implementation of policy that the government had approved in 2006. The brief titled: Applying Biotechnology in a Safe and Responsible Manner: Justification for the Bio-safety Law in Kenya and other literature on agricultural biotechnology were heavily quoted in the subsequent parliamentary debates. Recommendations were made for the creation of more awareness on biotechnology and ensuring that there was increased public participation in the process. In response to the recommendations made by the parliamentary committee on Agriculture, Lands and Natural Resources, the NCST, KARI, ISAAA AfriCentre and PBS organized a study tour and workshop on Bio-safety Bill 2008 for MPs in October 2008. The main objective of the tour was to expose the MPs to various institutional, technical and human capacities available in the country for responsible and safe research, development and application of modern agricultural biotechnologies. On realizing that some MPs were still opposed to the Bill even after sensitization workshops, the bill was table in parliament by an already revised and published Biosafety Bill 2008 which sailed through parliament. The president assented to the Bill two and a half months later, on 12 February 2009. The Bill was hence forth referred to as the Biosafety Act 2009.

Key ST&I Policy Issues in Kenya

It is pretty obvious that issues at hand are so much about inadequate machinery to coordinate focused national polices that support ST&I as a foundation for the Kenya Vision 2030. The current structures support much more of top-down policy development than bottom up approach as required by the stakeholders. Critical examination of how the ST&I and the Biosafety Bill were developed reveal several factors militating against the rational growth of ST&I in Kenya. The key factors are outlined below:

In order to realize the goals of Vision 2030, the Kenya Government must have a clear policy that discourages the top-down policy development process, but encourages a bottom-up approach in the ST&I policy development process. Such a policy would discourage experts from assuming that the views of difficult stakeholders had been captured. Quite often it is the less influential stakeholders who get assumed (assumed stakeholder participation), hence making the whole process to be top-down rather than bottom-up policy development process. As evidenced in the Biosafety Bill, the stakeholders who were driving up the process were institutional based and had a lot of influence in the corridors of power. Although there is a policy on public sector stakeholder partnerships, no clear mechanisms are available on how to engage all stakeholders in ST&I. This explains why at a certain stage during the development of the Bio-safety Bill, the disgruntled stakeholders almost derailed the process by organizing demonstrations and they even went to court to stop the Bill from progressing through Parliament. For PSSP policy to succeed, various stakeholder mapping tools need to be practiced. This will help the lead agencies or task-forces in assessing the levels of interest and power each stakeholder wields, and on how to engage each one of them at various stages of policy development. The Kenya Private Sector Alliance (KEPSA) which has been very vocal during the development of economic polices was missing in action during the development of ST&I and Bio-safety Bills. Keeping in mind that it is the private sector that benefits immensely from ST&I, there was need for the KEPSA to be fully represented in the stakeholder forums. The ST&I Bill lays emphasis on the integration of science and technology policies into the national economic development framework. However, there is need for the development of the stand-alone policies on various aspects of S&T matters *e.g.* nanotechnology, without being lumped together with the general science and technology policy. Such policies would provide for the recognition of the potential benefits from all technological innovations, while at the same time ensuring sustainable utilization and conservation of biological diversity, and minimizing the perceived risks associated with certain aspects or elements of the technology.

Without harmonized policies that support S&T research areas, there is the increasing danger for lack of coordinated research efforts by individual researchers in public universities and/or public institutions, as well as international research institutes. Much of the scientific activities in these institutions often reflect the interests of the concerned individual(s) or the particular institution(s), with very minimal (if any) information sharing among the stakeholders. This kind of scenario was evident during the development of the ST&I and Bio-safety Bills especially when it came to dealing with stakeholders. Most stakeholders were institutional based and this denied lay stakeholders from fully expressing their views. Any policy process that is being spearheaded by stakeholders who were receiving financial support from development partners was likely to create suspicion among other stakeholders. That explains why some stakeholders suspected that the institutional stakeholders were being lured by developed countries to dump GMOs in Kenya. Such a process is viewed as being influenced by foreign masters.

Inadequate institutional arrangement to foster ST&I policy development and implementation. It is evident from the foregoing, that there are several scattered ST&I activities within the country, whose existence are dictated upon by individual institutional research preferences and resources (which may not necessarily be guided by the national priorities). All national research institutes belong to sectoral ministries, but not the Ministry of Higher Education, Science and Technology (MOHEST). The research agenda in these institutions and even to a larger extend universities, is dictated by the availability of financial support from development partners. The current institutional setup is fraught with rigidity to collaboration in activities of potential mutual benefit to the stakeholders. Flexibility in sharing of research information, resources and experiences would ensure proper utilization of the current scarce resources. While there is a limited form of collaboration, one would want to see programs deliberately designed by interested parties to facilitate, for instance capacity building of Kenyan universities through the public and international research centres. Such institutions would provide the necessary support for studies in policy research. As things are now, ST&I policy studies are lacking in our training institutions. For that reasons Kenya is busy building her policy development human capital by sending

PhD students for further training abroad. The current set up is therefore an impediment when it comes to policy development. With stakeholders coming from institutions that hardly share information between themselves, it complicates the sharing of the same information during public debate. Proper guidelines on how to engage stakeholders in ST&I, need to be developed to avoid making the policy development process a political process, but rather a professional one. From the two Bills discussed, it was clear that the most important stakeholders were the experts forming institutional stakeholders and the politicians.

It is evident that Kenya lacks a translation mechanism for ST&I in Kenya. For the country to achieve the goals of Vision 2030, we need structures to help in translating evidence into tangible products and process innovations for Vision 2030. Apart from good polices as translation tools, there is great need to translate the technological research findings into tangible products, with the help of a strong engineering support system. The designs of research infrastructure should be in a manner that researchers should be exposed to the daily problems for which their research seeks solutions. This will work as a motivation factor to make them translate their findings into workable solutions. The pervasive and science-intensive nature of technology requires highly qualified and skilled human capital in various key disciplines (such as molecular biology, genetic engineering, biochemistry, chemistry and chemical engineering). There is also a great need to have researchers who are trained on regulatory/legal issues working side by side with laboratory researchers to prepare for a smooth and faster drafting of ST&I policies that might be needed for the upcoming technologies and innovations. Such skilled manpower in Kenya is seriously lacking, and is not likely to increase in the very near future due to the current training scenarios in Kenya. Most of the highly trained manpower usually receives their training in developed countries and upon their return, there is never support infrastructure to carry out research effectively. The current level of budgetary constraints in Kenya cannot sustain a rigorous training program in these foreign universities due to the increasing cost of education in these countries (For example, the Rockefeller Foundation puts the cost at US$ 100,000 per PhD student). Thus, the necessary rapid attainment of the critical mass required to drive technological process will be difficult in the foreseeable future (Avramovic, 1996). This dependency on foreign training is worrying and the Government needs to make a deliberate effort in supporting higher education adequately to facilitate research activities in public institutions.

Limited participation of indigenous firms in the industrial sector: The industrial sector is dominated by foreign firms that do not often solicit or require research input from local research establishments as they rely on researches conducted in their countries of origin. This could be the reason as to why KEPSA played a minimal role because there was less interest from its members. The situation is however, different in the tea, coffee, and horticultural sectors, research is heavily supported by public research institutions. This scenario does not auger well for the development of ST&I in Kenya, since there is evidence that commercial technologies have their roots in academia (Avramovic, M., 1996). Researchers in universities and biomedical research institutes carried out nearly all basic research from which high technologies were developed. Thus, private venture-capital companies, which could help, supply equity capital in support for development of local technologies is lacking. Besides any indigenous firm is very unlikely to invest their already limited capitalization or financial resources in long-gestation projects; particularly in a country where interest rates are often extremely high. As a result the research activities being carried out in Kenya is heavily funded by agencies from developed countries and therefore cannot fully answer out local policy needs. With regards to stakeholder engagement, the private sector stakeholders are always perceived as agents of foreign masters, who are hell-bent to dump foreign materials into the country.

Limited Funding in support of ST&I activities: In the 1979-1983 Development Plan, the Government adopted a long-term target to maintain a total national expenditure on research at a minimum level of 1 per cent of the Gross Domestic Product (GDP) (Government of Kenya, 1979). Most recently the government ratified the Abuja declaration, where it committed to start spending 1 per cent of its GDP on R&D. But this has not been forthcoming because of lack of reliable statistics on ST&I indicators. The bureaucrats have always argued that since the Government foots the bill for employees in all public research institutes and universities, then its expenditure on R&D could even be more than 1 per cent. There is an urgent need for the country to have reliable ST&I indicators to inform policies in the sector.

Lack of Public Awareness: The public is generally ill-informed about ST&I issues, and this includes most graduates of ST&I disciplines. The reason is that very little applied S&T is being taught at the university and tertiary colleges. The collaboration between universities and the industry is very weak and therefore graduates end up acquiring theories from universities. It is therefore critical that national institutions of higher learning address this as a matter of urgency. In order to have a working ST&I policy development process, it has been proposed that "connections with the public should no longer be viewed as exercises in better communications from a privileged elite" (Ozolina *et al.*, 2009). This re-emphasizes the point the need to connect with the public in the process of S and T at an earlier stage instead of being invited to support what has already been developed by the expert as in the case of the Bio-safety Bill.

Recommendations and Conclusions

The current institutional framework works as an impediment to the engagement of stakeholders in ST&I policy process. The fact that all public research institutes, universities and technical training institutes work under different legal mandates and fall in different ministries, contributes to stakeholder infighting during policy development. Universities and research institutes feel that they are better placed and equipped than other government agencies to spearhead the development of policies. This has in effect affected the policy process in which the ministries are the lead agency. For example, ST&I development in this country lacks a coordinated approach in solving national issues, because while all the science and technology policies are under NCST, its establishment cannot facilitate proper management of their programs. Therefore, in order for the NCST to be effective in propelling the policy formulation and implementation processes, radical changes are needed.

One of such changes would be the increased expertise and budget allocation for NCST and to the nation's R&D sector in general. The current budgetary distribution (over 60 per cent of their budgetary allocation is for salaries and rental fees and the rest is left for science and technology support activities) is absolutely inadequate and lopsided.

Another area where there is lack of direction is the Public Sector Stakeholder Engagement Policy (PSSP) that was developed by the Ministry of State for Public Service. Whereas the policy's intentions are good and clear, it falls short of giving adequate mechanisms for engaging the stakeholders especially in Science and Technology policy issues. There is a need to establish modern mechanisms for engaging the diverse groups of stakeholders in the country. To improve on infrastructure like, in establishing a national research institute in engineering and technology, there is need for involvement of the stakeholders in the private sector who are drawing huge benefits from engineering and technology.

NCST must be mandated to enter into partnerships with the African Center for Technology Studies (ACTS) and any other policy organizations as an advisor on policy analysis matters in S&T in general.

The idea of using upstream public engagement of the public in the policy process should be encouraged. The Demos report which proposes for dealing with nanotechnology can as well be relevant to the Kenya's situation. The report emphasizes the need "to articulate the visions, promises and expectations of the technology at an earlier stage, and make them the focal point of upstream public engagement" (Wilsdon and Willis, 2004).

Kenya should strive to enhance the relationship between science and the stakeholders or public. In this relationship, all stakeholders should be considered as equals. Unlike the previous frameworks, which gave science a position of power and perceived the public to be more or less deficit, the approach acknowledges that the parties affect each other (Reddy *et al.*, 2009b).

The stakeholder engagement should not just be limited to policy formulation and adoption, but should be extended to implementation as well. This approach is important for fostering a more inclusive government through public consultation as a way of consolidating democracy (Reddy *et al.*, 2009a). Implementation in this regard entails further action, which may lead to the establishment of regulatory frameworks in Nanotechnology. The approach is also a way of ensuring that public engagement is not exercised in a top-down manner.

Conclusion

Kenya's experience can certainly inform other countries. The lessons leant as discussed above are important for sealing the gap that tends to emerge in the creation of regulatory frameworks and implementing them. The alternative means for improving Kenya's ST&I policy development process, which is aimed at enhancing the relationship between science and the public, can go a long way in bridging the gaps that have been discussed in this paper.

References

ABSF, 2003. *Summary Proceedings of the Biotechnology and Bio-safety Workshop* held on 4th–6th August, Nairobi, Kenya.

ABSF, 2003. *Proceedings of the Biotechnology and Bio-safety Workshop* held on Oct. 31st Nov. 1st, Mombasa, Kenya.

Avramovic, M., 1996. An Affordable Development? In: *Biotechnology, Economics and the Implications for the Third World*, Chapter 2. Zed Books Ltd, London, UK.

Chopyak, J. and Levesque, P., 2002. Public participation in science and technology decision making: trends for the future. *Technology in Society*, 24: 155–166.

Economic Recovery Strategy for Wealth and Employment Creation, 2007. *Economics*, 19: 211–240.

Einsiedel, E.F., Jelsoe, E. and Breck, T., 2001. Publics at the technology table: the consensus conference in Denmark, Canada and Australia. *Public Understanding Sci.*, 10: 1–16.

Outram, S.M., 2010. Science Communication in Sub-Saharan Africa: The case of GMOs. *Journal of International Development*, 22: 341–351.

Ozolina, Z., Mitcham, C., Stilgoe, J., Andanda, P., Kaiser, M., Nielsen, L., Stehr, N. and Qiu, R.Z., 2009. *Global Governance of Science.* Report of the Expert Group on Global Governance of Science to the Science, Economy and Society Directorate, Directorate-General for Research, European Commission. Available at http://ec.europa.eu/research/science- society/document_library/pdf_06/global-governance-020609_en.pdf

Peltzman, S., 1976. Toward a more general theory of regulation. *Journal of Law and Economics*, 19: 211–240.

Public Sector Stakeholder Partnership Policy (PSSP), 2007. Ministry of State for Public Service, Government Printer, Nairobi.

Reddy, V., Juan, A., Gastrow, M. and Bantwini, B., 2009a. *Science and the Publics: A Review*, July.

Reddy, V., Juan, A., Gastrow, M. and Bantwini, B., 2009b. Public understanding of science. *HSRC Review*, 7(3) September.

Republic of Kenya, 1974. *Development Plan 1979–1983.* Government Printer, Nairobi.

Republic of Kenya, 1979. *Development Plan 1979–1983.* Government Printer, Nairobi.

Republic of Kenya, 1993. *Development Plan 1989-1993.* Government Printer, Nairobi.

Sclove, R., Scammell, M. and Holland, B., 1998. *Community-Based Research in the U.S.: An Introductory Reconnaissance.* Amherst (MA): The Loka Institute.

Stigler, G., 1971. The theory of economic regulation. *Bell Journal of Economics*, 2: 321.

Thitai, G., 2000. Biosafety regulations and structures in East Africa. Unpublished paper presented at a *Biotechnology Awareness Workshop*, Organized by ABSF in March, Nairobi, Kenya.

Wilsdon, J. and Willis, R., 2004. *See-through Science: Why Public Engagement Needs to Move Upstream.* DEMOS, London.

Abbreviations

AAS: African Academy of Sciences
AATF: African Agricultural Technology Foundation
ABSF: Biotechnology Stakeholders Forum
ABSF: African Biotechnology Stakeholders Forum
ACBF: African Capacity Building Foundation
ACTS: African Center for Technology Studies
AMREF: African Medical Research Foundation
ASALs: Arid and Semi Arid Lands
ASARECA: Association for Strengthening Agricultural Research in Eastern and Central Africa.
BICs: Biotechnology Information Centres
BioeARN: East African Regional Programmes and Research Network for Biotechnology, Biosafety and Biotechnology Policy Development
CBD: Convention on Biological Diversity
CBR: Community-based research
CDC: Centre for Disease Control
CNHR: Consortium for National Health Research
DBVD: Division of Vector bourn diseases
ECABIC: Eastern and Central Africa Biotechnology Information Centre
ERS: Economic Recovery Strategy
ERSEWC: Economic Recovery Strategy for Employment and Wealth Creation
GMO: Genetically Modified Organisms
IBC: Institutional Bio-safety Committees
ICIPE: International Centre for Insect Physiology and Ecology
ICRAF: International Council for Research in Agroforestry
ILRI: International Livestock Research Institute
IPRSP: Interim Poverty Reduction Strategy Paper
ISAAA: International Service for the Acquisition of Agri-biotech Applications
JICA: Japanese International Cooperation Agency
KARI: Kenya Agricultural Research Institute.
KBIC: Kenya Biotechnology Information Centre
KC: Knowledge Centre
KEBS: Kenya Bureau of Standards
KEFRI: Kenya Forest Research Institute
KEGCO: Kenya GMO Concern
KEMFRI: Kenya Marine and Fisheries Research Institute

KEMRI: Kenya Medical Research Institute
KENIA: Kenya National Innovation Agency
KENREF: Kenya National Research Fund
KEPHIS: Kenya Plant Health Inspectorate Service
KIPI: Kenya Intellectual Property Institute
KIPPRA: Kenya Institute for Public Policy Research and Analysis
KIRDI: Kenya Industrial Research Development Institute
KTDA: Kenya Tea Development Authority
MOHEST: Ministry of Higher Education Science and Technology
MPND: Ministry of Planning and National Development
MSFs: Ministerial Stakeholder Forums
NACC: National AIDS Control Council
NARC: National Rainbow Coalition
NBC: National Bio-safety Committee
NCST&I: National Commission for Science, Technology and Innovation
NCST: National Council for Science and Technology
NESC: National Economic and Social Council
PPA: Participatory Poverty Assessments
PRGF: Poverty Reduction and Growth Facility
PRI: Primate Research Institute
PRSPs: Poverty Reduction Strategy Papers
PSSPs: Public Sector Stakeholder Partnerships
ST&I: Science, Technology and Innovation
STAK: Seed Trade Association of Kenya
USAID: United States Assistance for International Development
WHO: World Heath Organization

Chapter 5

Overview of the Science, Technology and Innovations Institutional and Regulatory System in Malawi

Patrick J.J. Mphadzula

Principal Science and Technology Officer, Department of Science and Technology, Private Bag 328, Lilongwe 3. Malawi

Ministry of Education, Science and Technology

E-mail: mphadzula@yahoo.com, pmphadzula@gmail.com

ABSTRACT

The Malawi Government recognized the importance of Science, Technology and Innovations (STI) for socio-economic development as early as 1970s. The government therefore over the years established a number of institutions to promote and coordinate science and technology for sustainable socio-economic development of the country. Some of these institutions include the: National Research Council of Malawi (NRCM), Ministry of Research and Environmental Affairs (MOREA), Ministry of Science and Technology (MIST); and Department of Science and Technology. This paper, therefore, describes the institutional set up, legal and regulatory framework of science, technology and innovations in Malawi. The greatest challenge faced by the science and technology coordinating institutions was the instability of their existence. The other problem was duplication of efforts by the institutions, which resulted into some cases of conflicts and animosity. The Government resolved some of these problems by approving the 2002 Science and Technology Policy, and the Science and Technology Act 2003 which provided for the establishment of the National Commission for Science and Technology (NCST) and the small and lean Department of Science and Technology (DST). In this arrangement, the NCST is the implementing agency of the Science, Technology and Innovations Policy (promotes and coordinates science and technology activities) while the DST develops and reviews policies and strengthens international cooperation in Science, Technology and Innovations. It is envisaged that these institutional, legal and regulatory arrangements will put to rest all the problems faced in the past.

***Keywords**: Science, Technology, Innovations, Sustainable, Coordination, Regulate, Formulate and Review.*

Introduction

Malawi occupies an area of 11.8 million hectares, 9.4 million is land, the rest comprising of Lake Malawi and other smaller lakes. It has a population of 15.4 Million and a density of 102 persons per km² (NSO, 2008). Malawi is positioned in southeastern Africa, occupying a thin strip of land between Zambia and Mozambique. In the north it also shares a border with Tanzania (Figure 5.1).

Figure 5.1: Map of Malawi and its Neighbours (*Source*: Guide for Africa, 2010)

The Great Rift Valley traverses the country from north to south. In this deep trough lies Lake Malawi, the third-largest lake in Africa. The Shire River flows from the south end of the lake and joins the Zambezi River 400 km (249 miles) farther south in Mozambique. In the west, the land forms high plateaus, generally between 900 and 1,200 m (2,953 and 3,937 ft) above sea level. In the north, the Nyika Uplands rise as high as 2,600 m (8,530 ft). To the east of the Shire River the land rises to an elevation of 1,600 m (5,249 ft), and eventually to the country's highest point, Sapitwa Peak at 3,003 m (9,852 ft) near the Mozambique border (World Atlas, 2010).

In the extreme south, the elevation is just slightly above sea level. It is there that Malawi's lowest point sits at the junction of the Shire River and the border of Mozambique, 37 m (121 ft) above the sea level. The country has been landlocked until October, 2010 when a World inland port was officially opened to connect the country to Indian Ocean through Shire and Zambezi rivers. The country is endowed with relatively fertile land, extensive indigenous forests, and abundant fisheries. Agriculture dominates the economy, contributing 36 per cent of the GDP, 85 per cent to 90 per cent of the foreign earnings, and 85 per cent of the labour force. Most industries are agro-based, involved in low technology primary processing of agricultural raw materials and inputs. Low levels of technological development and a high dependency on imported raw materials characterize the light engineering industries (World Bank, 1992).

Malawi recognizes and appreciates that Science, Technology and Innovations (STI) are the drivers for socio-economic development. Countries in the South-East Asia, like Japan and Korea, have achieved high economic growth through the utilization of STI. In Africa, countries such as South Africa and the Maghreb region, have made greater socio-economic strides using STI. Malawi recognized the importance of science and technology for national socio-economic development as early as 1974, when it established the National Research Council of Malawi (NRCM) to coordinate the activities of Science and Technology in order to drive national development objectives and processes. Many institutional arrangements for coordination of Science and Technology have been put in place over the years. This paper, therefore, attempts to highlight the institutional set up, legal and regulatory framework of science and technology system in Malawi over the years.

Analysis of Science and Technology Institutional, Legal and Regulatory Framework for Malawi

The Malawi Government recognised the importance of Science and Technology in socio-economic development as early as the 1970s. The Government therefore over the years established a number of institutions to promote and coordinate science and technology for sustainable socio-economic development of the country. Some of these institutions include: National Research Council of Malawi (NRCM), Ministry of Research and Environment (MOREA), Ministry of Industry, Science and Technology (MIST); and Department of Science and Technology (DST).

National Research Council of Malawi

The National Research Council of Malawi (NRCM) was a first body to serve as a principal advisory organ of Government on all matters relating to research, science and technology. It was constituted in 1974 by a presidential decree (Malawi Gazette, 1976) and it was reporting to the Office of the President and Cabinet (OPC). The OPC was the right and most ideal office for the NRCM to report to because it is the highest office of the land of Malawi and has control of all other ministries and departments. It was then easy for the NRCM to coordinate the science and technology activities in various ministries and departments under the arm of the OPC. The mission of the NCRM was to promote and coordinate the development and application of research for the realisation of sustainable socio-economic development in line with the Malawi Government's Development Goal. It had a Secretariat and a Council which was chaired by the Chief Secretary for Public Service. The Council was operating through the tenets of subject specialists committees such as:

- Agricultural Sciences Committee,
- Committee on Scientific and Industrial Research and Development,
- Genetics Resources and Biotechnology Committee,
- Legal and Patenting Policies Committee,
- National Documentation and Information Coordinating Committee,
- Science Competitions Committee,
- National Health Science Research Committee, Building and Construction Research Committee and;
- Research Programmes Committee.

The membership of the above named committees was drawn from experts in their own rights. The majority of them were from the universities, research institutions, civil society, ministries and government departments. The NRCM faced many obstacles such as inadequate power to influence high budgetary considerations and the instability and uncertainty of its existence. As it has been mentioned earlier, the NRCM was established with the Secretariat in the office of President and Cabinet (OPC) in 1974, later, in 1988, the activities of the NRCM were merged with those of the Environmental Unit of the Ministry of Forestry and Natural Resources. In 1991, the NRCM Secretariat was elevated to a full department which was later elevated to a full Ministry called the Ministry of Research and Environmental Affairs (MOREA). In 1997, the MOREA was abolished and the NRCM was reverted back to the OPC. The NRCM was stable from 1997 to early 2004; and in mid 2004 the government of Malawi created a Ministry of Science and Technology (MIST) to work along with the NRCM (NRCM, 2002).

Ministry of Industry, Science and Technology (MIST) and the old Department of Science and Technology

The Ministry of Industry, Science and Technology (MIST) was created shortly after the May 2004 general elections. The articulated vision of MIST was to transform Malawi into an industrialized nation with scientifically and technologically led sustainable growth and development. Its mandate was to promote industrial development through the application of S &T, and the ministry consisted of two technical departments, namely Science and Technology, and Industry. However, the

MIST was dissolved in 2006, with the Industry Department residing in the then created Ministry of Industry, Trade and Private Sector Development. While the control of the Department of Science and Technology of the MIST moved back to the Office of the President and Cabinet. This meant that the department was directly reporting to the president just as the NRCM was also reporting to the same (SciDevNet, 2006).

The mandate of the Department of Science and Technology was to regulate, support, promote and coordinate the development and application of Science and Technology in order to create wealth and improve the quality of life. The Department had four corporate objectives (portal.unesco.org, 2006). These were to:

- Initiate the promotion and maintenance of the regulatory framework of S and T for wealth creation;
- Establish an effective capacity building framework for efficient development, application and transfer of Science and Technology;
- Develop and commercialise Technology; and
- Enhance, promote, and popularise Science and Technology

Some of the objectives of this Department were similar to those of the NRCM and a conflict between the two institutions was born. In May 2007, the Malawian president merged Education, and Science and Technology under a single ministry, of which he personally took charge. However the merge did not solve the conflict that existed between the Department of Science and Technology and the NRCM. The instability and conflicts among institutions mandated to coordinate science and technology activities were realised as early as 1990. These problems influenced the authorities to formulate the Science and Technology Policy. Malawi's first Science and Technology policy was formulated and adopted in 1991. However, the 1991 policy did not explicitly integrate Science and Technology issues into national development planning. This weakness necessitated the Government to review the policy and to adopt a new Science and Technology policy in 2002. The revised policy contained clear-cut Science and Technology development objectives and strategies for all priority sectors of the Malawi economy. The policy also articulated the Government's commitment to increase funding for S&T as well to strengthen the institutional and legal framework for S&T. In terms of the institutional capacity, the policy provided for the establishment of the National Commission for Science and Technology (NCST) as a way of resolving the problems of the implementation arrangements faced in the Science and Technology sector (NRCM, 2002).

In late 2003, a Science and Technology Act was introduced to back up the 2002 Science and Technology Policy (Malawi Government, 2003).

National Commission for Science and Technology (NCST)

The National Commission for Science and Technology (NCST) became operational as a regulating and coordinating agency for all research in Malawi in December 2009. NCST was created from a merger of the National Research Council of Malawi and the Department of Science and Technology of the Ministry of Science and Technology (MIST) as called for by the Science and Technology Act of 2003

(HRCSI Newsletter, April, 2010). The Commission's function is to advise the government and other stakeholders on all S&T matters in order to achieve a science and technology-led development (Malawi Government, 2003). Among its various responsibilities, the Commission will also:

- ✰ Chart out national directions and establish national priorities in S and T development in relation to socio-economic development needs;
- ✰ Source funding from within and outside Malawi to finance the national R&D effort and allocate funds to institutions based on set priorities;
- ✰ Promote and advocate for the development of S and T human resources by building capacity in S& T education and training programmes;
- ✰ Encourage the establishment and promote the co-ordination of research institutions that undertake R&D activities;
- ✰ Develop and synthesize S and T indicators;
- ✰ Prepare a biennial report on the state of Malawian S and T for the National Assembly.

The Commission will have various important committee structures (NRCM, 2002), such as:

- ✰ A Parliamentary Committee responsible for Science and Technology, which will provide a voice for S and T matters in the National Assembly. The Committee will ensure that S and T is integrated into the national budget by the time it is approved by Parliament and that S and T is integrated in the socio-economic development processes of the country.
- ✰ A Cabinet committee responsible for Science and Technology, which will monitor the development and application of Science and Technology in national development processes.
- ✰ Sectoral Committees, which will be composed of S and T directors from the relevant sectoral ministries, as a link between the sectoral ministry and the Commission. The overall responsibility of the S and T directors will be to co-ordinate S and T issues in their relevant ministries.

New Department of Science and Technology

The National Commission for Science and Technology was established as a Statutory Corporation and not as a full government department. The authorities therefore decided to reconstitute a small and lean Department of Science and Technology under the Ministry of Education, Science and Technology to handle issues of policies and international cooperation in Science and Technology. It was realized that these issues could not be properly handled by a statutory body such as the Commission. The functions of this Department of Science and Technology are to:

- ✰ Formulate, review and publicise policies in Science, Technology and Innovations
- ✰ Establish and strengthen the local and international linkages in Science, Technology and Innovations.

The Department of Science and Technology reports to the Ministry of Education, Science and Technology and it operates with two Divisions handling policy issues and international cooperation as shown in the structure in Figure 5.2.

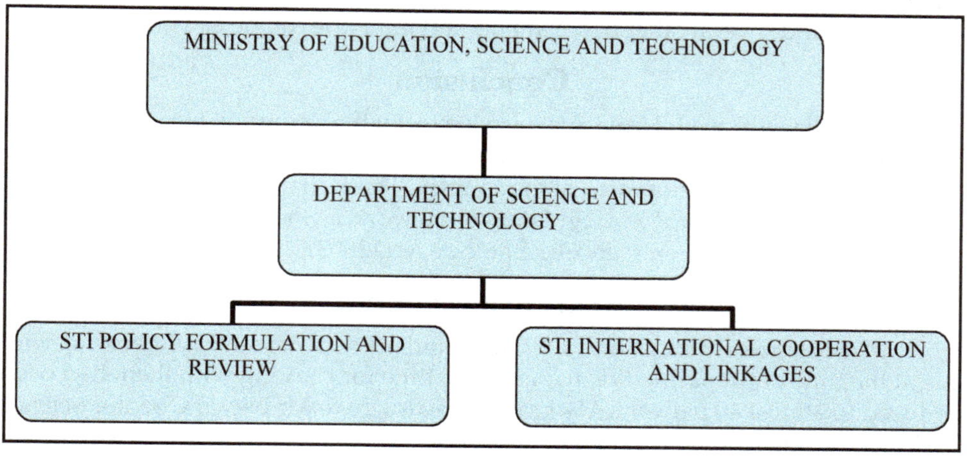

Figure 5.2: Structure of the Department of Science and Technology

The Department of Science and Technology is in the process of reviewing the Science and Technology Policy, 2002 in order to include an implementation plan and to align it to the Malawi Growth and Development strategy. Furthermore, the reviewed policy will provide a direction on international science, technology and innovation instruments to which Malawi is a party. The reviewed policy will be linked to Vision 2020, a long-term national development strategy, which envisages Malawi as a technologically driven middle income economy. It will also be clearly linked to a number of sector-specific policies such as National Environmental Policy (2004), Food Security Policy (2008), National Youth Policy (1999), National Biotechnology and Bio-safety Policy (2008), National Irrigation Policy (2000), National Sanitation Policy (2006) and the National Nutrition Policy (2008). The Department is also drafting the Intellectual Property (IP) Policy and has so far submitted it to Cabinet for approval.

Discussion

The Science, Technology and Innovations system in Malawi has been weak for a number of years, due to the absence of a legal and regulatory framework. Science and Technology coordinating institutions were formed at an ad hoc basis. For example, the NRCM was formed following a presidential decree. There was no regulation and legal basis to support its existence. The NRCM was supposed to be regulated by an Act of Parliament. The Ministry of Science and Technology, which was created in 2004, did not last long as all ministries are under the prerogative of the Head of State. In some cases conflicts and duplication of efforts existed as the Government was not conducting needs assessments before creating the institutions. The Government of Malawi got it right when it approved the Science and Technology Policy in 2002 in order to resolve the problems. The approval of the policy paved way for the introduction of the Science and Technology Act (2003). The Science and

Technology Act (2003) has created an enabling environment for the National Commission for Science and Technology (NCST) and the Department of Science and Technology to be established and work together. The two institutions target the same stakeholders such as universities, agricultural stations, health-related research institutions and several other departments and ministries of the Government.

Conclusion

The challenges faced should serve as a lesson for the future and tolerance between institutions is paramount for the advancement application of science and technology for socio-economic development of the country. These challenges have been solved by putting in place the legal and regulatory framework, *i.e.* the Science and Technology Policy (2002) and the Science and Technology Act (2003).

Acknowledgement

I wish to sincerely thank my wife Gloria and my son Mayamiko for bearing with me at the time of preparing this paper as the time for chatting with them had been reduced to almost 10 per cent. My gratitudes also go to Mr Ngwira, Senior Science and Technology Officer, Department of Science and Technology, who provided moral and technical support on the paper. Let me also thank NAM for sponsoring my trip to Iran to attend and present the paper at the International Workshop on Science and Technology Policy and Sustainable Development in Tehran, ISLAMIC REPUBLIC OF IRAN, 5-7 January, 2011. Lastly, let God Almighty be praised for the precious life and energy during the preparation of the paper.

References

Guide for Africa, 2010. *Map of Malawi*. http://www.guideforafrica.com/malawi/page/3.

Health Research Capacity Strengthening Initiative (HRCSI), Newsletter, April, 2010.

Malawi Gazette, 1976. General Notice No. 398.

Malawi Government, 2003. *Science and Technology Act 2003.*

National Research Council of Malawi (NRCM), 2002. *Science and Technology Policy.*

National Statistic Office (NSO), 2008. *The Population and Housing Census.*

SciDevNet, 2006. *Malawi: President takes Charge of Science.*

World Atlas, 2010. *Map of Malawi.* www.worldatlas.com/webimage/countrys/africa/mw.

World Bank, 1992. *World Development Report*, 1992.

www.unesco.org, 2006. *The Science and Technology Profile of the Republic of Malawi.*

Chapter 6

Science and Technology Policy and Sustainable Development in Myanmar

Myint Thein

Director,

Material Science and Material Engineering Research Department,

Ministry of Science and Technology of Myanmar,

Kyaukse, Myanmar

E-mail: myinttheinssmaeu@gmail.com

ABSTRACT

In this paper, Innovative Practices in Science and Technology towards Education for Sustainable Development of Myanmar are presented. Myanmar is improving the infrastructure in various fields for equal development of all states and regions of the union. Among several developments, the sustainable development of education is highlighted for the year 2010 in this paper. The Ministry of Science and Technology has been making efforts with accelerated momentum to achieve advancement in science and engineering in the country and to obtain the technological expertise essential for the country to become a modern and developed one. The Ministry of Science and Technology has already opened 29 Technological Universities, 25 Computer Universities, 5 Government Technical Colleges, 12 Government Technical Institutes (GTI) and 30 Government Technical High Schools (GTHS) in this year in order to improve education of Myanmar. To build a modern developed nation, not only advanced technological education, but also vocational education system needs to produce more and more high-skilled and employable operators and technicians. This paper reports on the science and technology policy towards education for sustainable development in Myanmar.

Keywords*: Science, Technology, Development, Education, Policy, Infrastructure.*

Introduction

Today is the Age of knowledge, so human societies make good use of technological advancement. The development of technological and vocational education reflects the development of National Economy. The Ministry of Science and Technology comprises of six main Departments. Among them, the following two departments are conducting undergraduate and graduate courses on various fields in Technology and Computer Studies:

1. Department of Advanced Science and Technology
2. Department of Technical and Vocational Education

The locations of the Technological Universities (TU), Government Technical Institutes (GTI), Technical Training Schools (TTS) and Government Technical High Schools (GTHS) across the nation are shown in Figure 6.1 (Theingi, 2009).

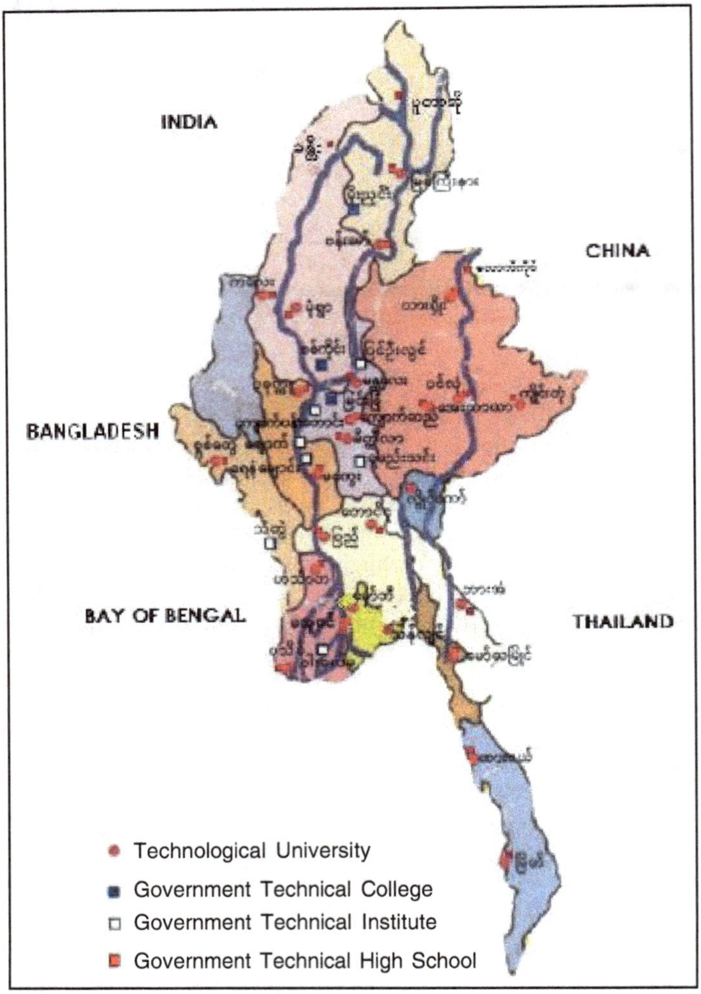

Figure 6.1: Technological Universities/Colleges/Institutes/High Schools

The aim of newly opened technological universities across the whole country is for every qualified citizen to be able to learn and study the advanced science and technologies in their native regions and then to able to improve industrial sectors in their regions.

The Practices on Science and Technology Policy in Myanmar

In order for every qualified citizen to be able to learn and study the advanced sciences and technologies, the Ministry of Science and Technology, until March 2010, has established the following new Universities, Colleges and Institutes (Zar Zar Wint, 2009):

1. Mandalay Technological University
2. Pyay Technological University
3. West Yangon Technological University
4. University of Computer Studies (Mandalay)
5. Aerospace and Aeronautical Engineering University
6. 29 Technological Universities
7. 25 Computer Universities
8. 5 Government Technical Colleges
9. 12 Government Technical Institutes
10. 30 Government Technical High Schools.

Not only the regular courses that have been available before the establishment of Ministry of Science and Technology, but also new advanced undergraduate and post-graduate courses, become available now at the Technological Universities and Colleges under the Ministry of Science and Technology. The newly established Universities, Colleges and Institutes provide increased capacity for training more students and trainees.

The Ministry of Science and Technology has sent its staff and qualified students in Engineering and Computer Studies to foreign countries for further studies, training, and for attending international seminars and conferences.

In order to modernize the Technological Universities, Technical Colleges, and Institutes, Universities of Computer Studies, and Computer Colleges in Myanmar so that they would reach international standards, modern advanced laboratory equipment for research work and materials for teaching aids have been bought and supplied. The budget for buying and supplying advanced laboratory equipment, teaching aids, and machines for research projects to Universities, Colleges and Institutes under the Ministry of Science and Technology is increased year by year.

For the experimental works of the technological students, the Ministry of Science and Technology has bought and installed modern advanced workshop machines including the Computer Numerical Control (CNC) machines, such as CNC Lathe, CNC Milling Machine, CNC Machining Centre, Laboratory equipment and Research facilities.

Research and Development Programs

At Master's and Doctoral Research Programs, emphasis is placed on those projects that will be useful in rural areas development and industrial development of the nation. Some of the research programs that are being carried out are the following:

For the development of rural areas, the following research works are being carried out under the close supervision of the Minister for the Ministry of Science and Technology:

1. Research on the production of natural fertilizers, liquid fish fertilizers, and bio-fertilizers
2. Research on Tissue Culture
3. Research on supplying power to households in the Country Site where power distribution from the national grid is not possible by using biogas
4. Research on Construction of gasifiers which utilize rice chaff and those which utilize woodchips
5. Research on the production of bio-diesel
6. Research on the manufacture of 22 H.P. engine for general applications in agriculture.

For the purpose of industrial development, the following research activities are being carried out under the supervision of the Minister for the Ministry of Science and Technology:

1. Research on Modeling and Construction of Computer Numerical Control Machine
2. Research on Modeling and Construction of Engines of Medium Size and Gear Box
3. Research on Modeling and Construction of Electronic devices used in Industry
4. Research on Modeling and Construction of Induction Furnace and Induction Hardening Machine.

The following research and development activities are being carried out at the Universities of Computer Studies under the Ministry of Science and Technology:

1. Translation from English to Myanmar language using Myanmar Unicode by Natural Language Processing (NLP) Project
2. Setting up E-learning centres, E-resource centres, E-library centres and Computer training centres at universities and colleges under MOST
3. Research on electronic data broadcasting system
4. Research on video-conferencing system

The Ministry of Science and Technology signed agreements with the following countries for science and technological collaboration:

1. People's Republic of China

2. Republic of India
3. Republic of Korea
4. Islamic Republic of Pakistan

The Ministry of Science and Technology has been making great efforts to raise the education standard of the nation and has been able to provide highly-qualified graduates in Technology and Computer Studies by:

- ☆ Nurturing qualified and competent teachers
- ☆ Making the technical education level of students in urban and rural areas equivalent by sending qualified and competent teaching staff to universities and colleges throughout the country
- ☆ Purchasing laboratory equipment, teaching aids, and other research facilities
- ☆ Carrying out research and development projects.

As a result of the above efforts, highly-qualified students and teachers of Technology and Computer Studies produced by the Ministry of Science and Technology have been able to carry out research in their relevant specialized fields successfully. They have been able to submit technical papers on their research results to International Journals and International Conferences. The papers have been accepted for publication, as well as presented in international conferences.

Conclusion

With the capacity of qualified and competent teachers we have, utilizing modern academic course syllabus and curriculum, advanced teaching aids, and carrying out research and development activities, the Ministry of Science and Technology has been making efforts with accelerated momentum to achieve advancement in science and engineering of the country, and to obtain the technological expertise essential for the country to become a modern and developed one. The Ministry will continue its efforts in human resources development activities, as well as efforts in carrying out research systematically and successfully on rural area development, industrial development, defense, health and education in the nation.

References

Theingi, 2009. *Reorienting TVET Policy towards Education for Sustainable Development in Myanmar.*

Zar Zar Wint, 2009. *Harnessing Regional Qualifications Framework to Integrate Quality Assurance in TVET (Myanmar).*

Chapter 7

Development of Science and Technology Policy in Sri Lanka

Nalini B. Ratnasiri

National Science and Technology Commission, Sri Lanka

E-mail: chnastec@sltnet.lk, jratnasiri@sltnet.lk, chnastec@yahoo.com

ABSTRACT

Scientific and technological innovations are important for a country to improve its competitiveness and productivity to achieve a higher standard of living and a better quality of life for its people. It is therefore, essential that a strong commitment be made to harness the potential of S&T to ensure economic development and the welfare of the people. The need for a comprehensive and consistently implemented national policy on S&T has been long felt in Sri Lanka. The National Science and Technology Commission of Sri Lanka developed the science and technology policy of the country after a process of wide consultation with all stakeholders. The government in 2009 approved this policy as the National Science and Technology Policy. This is an overarching policy focusing on two main aspects, the development of S&T in the country and the application of S&T for national development. Under the development of S&T, the areas covered are: S&T Innovation, Capabilities of S&T for National Development, Human Resource Base for S&T activities, Research and Development and Quality and Performance of S&T institutions. Under application of S&T for national development the main areas covered are Transfer of technologies, Sustainable use of natural resources, Promoting of indigenous knowledge, Innovations and Intellectual Property Rights and S&T and human society. The policy comprises ten objectives, one each on the areas stated above. There are varying numbers of strategies under each objective. The policy also proposes a number of initiatives under each strategy and the various agencies and institutions responsible for their implementation.

Keywords: *Sri Lanka, National Science and Technology Policy.*

Introduction

Sri Lanka is an island nation located south east of the Indian Subcontinent. It covers a total area of 65,610 km^2 and had a population of 20.45 million in 2009. Sri Lanka's GDP (at current price) in 2009 has been US$ 42 billion, with an average annual real growth of 6.0 per cent during the period 2004-2008, and a per capita GDP of US$ 2,053. The main contributor to GDP has been the services sector (59 per cent), followed by the industrial sector (29 per cent) and the agriculture sector (12 per cent).

According to the National Science Foundation (NSF) Statistical Handbook on ST&I (2008), Sri Lanka has spent US$ 46.1 million on R&D amounting to 0.11 per cent of the GDP. The annual investment by the state sector has amounted to 0.08 per cent of the GDP while the investment from the private sector has been only 0.02 per cent of the GDP. A share of 40 per cent of the R&D expenditure has been on applied research, 31 per cent on basic research and 29 per cent on experimental development research. The number of scientists engaged in R&D has been 4,037 with a full time equivalent (FTE) of 1,972. The patent office of Sri Lanka has registered 168 patents and the number of articles published by Sri Lankan scientists in the SCI Journals has amounted to 303. Sri Lanka's Gross domestic Expenditure on Research and Development (GERD), a key indicator has been in the range 0.15-0.2 per cent of the GDP, which is low by international standards and had therefore failed to facilitate the desired development in S&T capability.

When looking back to the history of Sri Lanka which dates back to several centuries BC, it has been observed by historians and archeologists that the country has had an advanced civilization evidenced by many archeological remains and the unique ancient irrigation system developed for water management. The system consisted of a large number of reservoirs built in the water-deficit dry zone, which were interlinked by canals built with indigenous advanced hydraulic technology. These tanks and canals are in use even today and is a good example of the application of technology for sustainable development, and in today's context adaptation to climate change. This is evidence of the high technological capability of the country during that period.

History of S&T Policy Development in Sri Lanka

The earliest attempt to formulate an S&T Policy in Sri Lanka was made by the Ceylon Association for the Advancement of Science during the Period 1950 – 1960. Following the formation of the National Science Council (NSC) in 1969, there is evidence that the first seven point S&T policy statement was developed by the NSC along with the Sri Lanka Association for the Advancement of Science (SLAAS) in 1978. His Excellency the President of the Democratic Socialist Republic of Sri Lanka, inaugurating the annual sessions of the SLAAS in 1978, made the following seven-point policy statement on Science and Technology.

1. Use Science as an integral part of the development strategy of the country and involve scientists in the formulation of policy and in decision making at the highest levels.
2. Foster scientific activity in all its aspects and in its widest possible scope, and to maintain a vigorous drive towards self reliance in National Scientific and Technological Capacity.

3. Provide equal and adequate opportunities for all to acquire a basic education in science.
4. Ensure that the institutions of education and research produce scientists and technologists of the highest Caliber.
5. Provide scientists and technologists with good working conditions, adequate remuneration, due reorganization for their efforts and access to scientific knowledge and activities in other parts of the world.
6. Make available as widely as possible within the country the achievements of scientific and technical activities.
7. Cultivate among people, an appreciation of the value of science and scientific method as an indispensable part of modern society.

In December 1986, a Presidential Task Force (PTF) on S&T Development was appointed to look into all aspects of Science and Technology in the country, and in particular the implementation of the 7 policy objectives.

In 1991, this Task Force prepared and presented a report which consisted of a series of projects and programmes (PTF, 1991). Based on the recommendations of this PTF report, a Science and Technology Development Bill was presented to the parliament in 1994, which was unanimously passed as the Science and Technology Development Act No 11 of 1994. The new Act provided for the establishment of an apex body for policy development and brought all S&T institutions which were under the purview of the Ministry of Science and Technology under this Act.

In 1996, the successor to the NSC, the Natural Resources, Energy and Science Authority (NARESA) and subsequently the NSF (successor to NARESA) developed the following 10 policy elements.

1. The provision of adequate funds for S&T and R&D to be free from bureaucratic impediments.
2. A clearly defined National Development plan where Science and Technology planning is integrated with National Development at all levels.
3. Strengthen and upgrade S&T infrastructure focusing on existing institutions.
4. Provide better remuneration, reward systems and facilities to scientists.
5. Establish a process to develop skilled S&T manpower commencing with revision of school curricula.
6. Establish greater cooperation between S&T Institutions, Universities, and Industries for collaborative research that will have impact on National Development.
7. Selectively acquire and adapt scientific knowledge and technology from abroad.
8. Build a national S&T capability to develop local resource-based industries.
9. Establish a science culture among the work force, administrators, policy makers and all sections of the community.

10. Establish a national monitoring system to evaluate, modify and upgrade the S&T system and assess its impact on National Development.

Development of the Current National S&T Policy

In 1998, a new institution, The National Science and Technology Commission (NASTEC), was established as provided for in the 1994 Science and Technology Development Act, with the mandate for the development of policies and plans for the advancement of Science and Technology and the application of Science and Technology for national development. NASTEC initiated a programme to develop a complete and comprehensive S&T policy based on the previous draft policy documents and the new thinking necessitated by the environmental, social and economic factors for the sustainable development of the country, as well as the rapid advances and changes in global S&T scenario.

In 2007, NASTEC commenced a process of extensive consultation with relevant stakeholders to prepare the new S&T policy. The first draft was prepared by NASTEC, taking into consideration all previous documents and informal discussions held with stakeholders. This preliminary draft was made available to the Senior Scientists' Forum of NASTEC for their comments. A summary of the draft document was published in the print media inviting public comments. The full document was also posted in NASTEC website.

The views of members of the Senior Scientists' Forum and the general public were incorporated into the draft and a revised document was prepared which included the policy objectives and the relevant strategies. At this stage, NASTEC appointed five expert committees comprising eminent scientists, educationalists, representatives of learned societies, industrialists, researchers and other stakeholders. These committees, along with two consultants appointed to coordinate the work, studied the revised policy document, particularly its objectives and strategies, and suggested further revisions to the draft.

NASTEC is also mandated to organize a Biennial Conference on Science and Technology on a theme of topical interest, and in 2008, the Conference theme was Science and Technology Policy Development. Over 150 senior scientists, technologists and other stakeholders, who participated at this conference, critically reviewed the draft policy proposing many more revisions. The final revised policy document incorporating all suggestions received was prepared by the NASTEC. The process of policy development is illustrated in Figure 7.1.

The final draft of the policy document was presented to the Minister of Science and Technology in early 2009, who submitted it to the Cabinet of Ministers for its approval. The Cabinet approved document was published in 2009 as the National Science and Technology Policy (NSTP) of Sri Lanka (NASTEC, 2009).

Key Elements of the Policy

The National Science and Technology Policy deals with two major aspects:

1. The development of S&T in the country
2. Application of S&T for national development

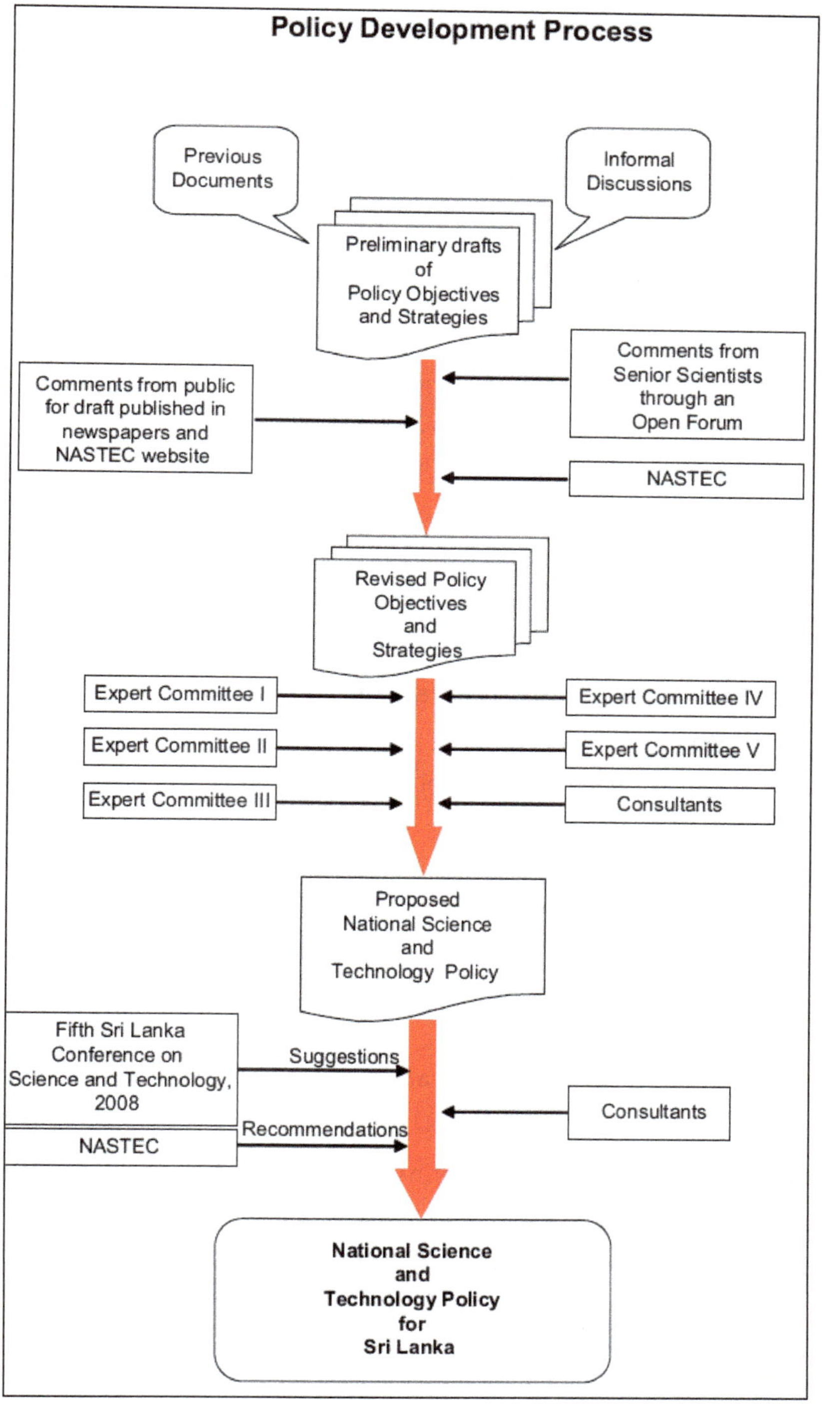

Figure 7.1: The Development Process of the Sri Lanka's National Science and Technology Policy

Under the aspect of S&T development, the policy focuses on the following:

1. S&T Innovation
2. Capabilities of S&T for national development
3. Human resource base for S&T activities
4. Research and development
5. Quality and performance of S&T institutions.

Under the aspect of S&T application for national development, the policy focuses on the following:

1. Transfer of technologies
2. Sustainable use of natural resources.
3. Promoting of indigenous knowledge
4. Innovations and Intellectual Property Rights
5. S&T and human society.

The policy document details out a number of strategies under each policy objective and these are given below.

Policy Objectives and Strategies

Science, Technology and Innovation Culture

Policy Objective

Foster a science, technology and innovation culture that effectively reaches all citizens of the country.

Strategies

1. Provide equal and adequate opportunities for all to acquire a basic science education.
2. Encourage among students an inquiring mind and the application of scientific methodologies for efficiency and productivity in every day life.
3. Promote an appreciation of Science and Technology among the public, leading to a culture of innovation and entrepreneurship, as an essential aspect of a progressive society.

Capability in Science and Technology for National Development

Policy Objective

Enhance Science and Technology capability for national development, make use of Science and Technology expertise in the national planning process, and strengthen governance and policy implementation mechanisms.

Strategies

1. Progressively increase the investment for Science and Technology up to 1.5 per cent of GDP by the year 2016, with the public sector contribution being at least 1 per cent.

2. Develop and strengthen the existing Science and Technology institutions and universities to generate quality research and train scientists.
3. Improve the autonomy and flexibility of Science and Technology institutions.
4. Promote strong linkages with global Science and Technology initiatives and with international centres of excellence in collaborative Research and Development and technology development.
5. Include scientists and technologists in the formulation of national development policies and plans, review strategies, legislation, and in decision-making and implementation, at the national and provincial levels, properly utilizing relevant scientific data.
6. Establish an Inter-Ministerial Committee with the Head of State as the Chairman and the Minister of Science and Technology as Vice Chairman to direct, coordinate and monitor the implementation of Science and Technology Policy. The Ministry of Science and Technology shall be the executive arm of the committee.

Human Resource Base

Policy Objective

Build up, and progressively expand and improve the resource base of scientists and technologists necessary to respond to the developmental needs of the country.

Strategies

1. Increase the number of researchers to reach the world average and develop a critical mass of scientists and technologists trained, particularly in advanced technologies, to effectively support the socio-economic needs of the country.
2. Increase the number of engineering graduates and attract engineers to Research and Development while ensuring opportunities for their advanced training with international collaboration, and support from the public and private sectors.
3. Provide opportunities for scientists and technologists to acquire advanced knowledge in research and international practices in Science and Technology.
4. Establish a National Research Cadre of high caliber scientists and technologists including Sri Lankan expatriate scientists, based on international criteria, ensuring suitable incentives to enable cutting-edge research.
5. Improve the remuneration systems, recognition and incentive schemes, and the working conditions of scientists and technologists in Science and Technology institutions.
6. Ensure opportunities for all segments of the population for vocational and tertiary education in Science and Technology irrespective of gender,

language, economic status or similar considerations and attract them to the Science and Technology field for their full and equal participation.

Research and Development

Policy Objective

Promote basic, applied and developmental research, particularly in areas of national importance and priority.

Strategies

1. Give priority to Research and Development in water, food, energy and environment and in the fields of nanotechnology and biotechnology, as well as information and communication technology, electronics, advanced materials and mechatronics.
2. Establish world class research centres to carry out cutting edge research in areas important for national development.
3. Improve facilities including state of the art ICT, for access to current scientific literature for all researchers and for effective networking.
4. Encourage collaborative research partnerships between Science and Technology institutions and industries.
5. Ensure that safe and ethical practices are observed in all Research and Development activities.
6. Establish a national research system under the Ministry of Science and Technology to coordinate, monitor and evaluate state funded Research and Development activities to ensure that they achieve the desired socio-economic benefits.

Technology Transfer

Policy Objective

Develop, or acquire and adapt, scientific knowledge and technologies for transfer to achieve progressive modernization of all sectors and to enhance the country's competitiveness in the world economy.

Strategies

1. Facilitate scaling-up of research based innovative processes and technologies to pilot and commercial scales.
2. Encourage industries and Research and Development institutions to give greater emphasis to high-tech innovations, technology transfer and commercialization.
3. Develop appropriate technologies suitable for transferring to small and medium enterprises, particularly in rural areas, through collaboration among Research and Development institutions, the SME sector and other stakeholders using mechanisms such as *Vidatha* Resource Centres.

4. Encourage international collaboration in Research and Development activities and joint ventures, for cost effective and rapid transfer of modern technologies, while ensuring adequate controls through well developed agreements for shared benefits.
5. Facilitate entrepreneurship and foresight activities among scientists, technologists, researchers and inventors.

Natural Resources and the Environment

Policy Objective

Ensure sustainable use of natural resources for development while protecting the environment.

Strategies

1. Promote research related to conservation and sustainable use of biodiversity, and mineral, marine and other natural resources.
2. Strengthen the scientific capability for effective implementation of laws and regulations to protect natural resources and the environment.
3. Promote cleaner production technologies.

Indigenous Knowledge

Policy Objective

Document, research into the scientific basis of, and promote indigenous knowledge based technologies

Strategies

1. Retrieve, collate and document available indigenous knowledge and practices.
2. Research into the scientific basis of the available indigenous knowledge and technologies.
3. Ensure further development of traditional technologies to promote their application for demand driven value addition while ensuring Intellectual Property Rights.

Innovations and Intellectual Property Rights

Policy Objective

Develop a culture of innovation and Intellectual Property and ensure the protection of Intellectual Property Rights (IPR).

Strategies

1. Inculcate IP awareness among scientists, technologists, science and technology teachers and students and develop an IP culture in Research and Development institutions, universities, technical colleges and similar institutions.

2. Develop institutional policies and guidelines regarding innovations and IPR.
3. Establish mechanisms to facilitate scientists, technologists and inventors to patent their innovations.
4. Establish suitable mechanisms to effectively exploit innovations.

Quality and Performance of Science and Technology Institutions

Policy Objective

Ensure quality standards of Science and Technology Institutions, products and services to achieve national and international recognition

Strategies

1. Ensure the effectiveness of activities of the Science and Technology institutions to maintain quality standards of institutional products and services by establishing, implementing and maintaining internationally recognized quality management systems with periodical national and international review.
2. Promote Science and Technology institutions to achieve certification of management systems and laboratory accreditation status wherever applicable as per internationally recognized standards.
3. Ensure the effectiveness of certification and accreditation bodies and processes.

Science and Technology, and Human Security

Policy Objective

Promote the application of Science and Technology for human welfare, disaster management, adaptation to climate change, law enforcement and defence, to ensure human and national security.

Strategies

1. Use Science and Technology inputs to ensure security in water, food, shelter, energy, health and wellbeing.
2. Promote research on causal factors and effects of natural and manmade hazards to support mitigation and management of disasters.
3. Develop research programmes to address mitigation, vulnerability and adaptation aspects in respect of climate change effects.
4. Ensure security from crime through the application of Science and Technology methodologies.
5. Ensure national defence capability through research and the application of modern Science and Technology interventions.

The policy also proposes certain initiatives under each strategy and has identified the lead agency/institution responsible for the implementation of each of the strategies/initiatives. These are described in the full policy document available in the Commission's website www.nastec.lk.

The 10 policy objectives constitute an overarching statement that provides a framework for more specific policies and implementation plans to be developed by relevant authorities. A National Biotechnology Policy is already in place and a National Policy for Nano-Technology and Research are being prepared.

Policy Implementation

The policy recommends the establishment of an Inter-Ministerial Committee chaired by the Head of State with the Minister in charge of S&T as the Vice Chairman, for the purpose of coordinating S&T policy implementation through various ministries. The Committee is also expected to set funding priorities, decide on major R&D initiatives and monitor the implementation of the policy. It is proposed that the Ministry in charge of S&T will function as the executive arm of the Committee.

The 2009 NSTP will provide a long term framework for the growth and development of S&T in Sri Lanka, as well as application of S&T in national development, which will contribute to the achievement of the status of a developed nation in the foreseeable future. The vision of this policy is "A prosperous nation of scientifically literate and innovative people, with a strong and stable economy, based on highly developed scientific and technological capabilities".

Conclusion

Sri Lanka's NSTP was developed initially with inputs from experts and wide groups of stakeholders, including the general public. In the next phase, these ideas and views were presented to several area specific working groups, consisting of expert scientists and technologists in relevant fields, social scientists, economists, administrators and decision makers. These working groups subsequently gathered evidence, consider them and drafted reports and recommendations. These reports, recommendations and policies were then reviewed, revised and presented to the relevant authorities for approval and implementation.

References

NSF, 2010. *Sri Lanka Science, Technology and Innovation Statistical Handbook 2008.* National Science Foundation, Colombo

NASTEC, 2009. *National Science and Technology Policy.* National Science and Technology Commission, Ministry of Science and Technology, Colombo.

PTF, 1991. *Final Report of the Presidential Task Force on Science Technology Development* submitted to His Excellency the President of the Democratic Socialist Republic of Sri Lanka, Colombo.

Chapter 8

Evolution of Science and Technology Policy for Developing an Innovation Ecosystem and Promotion of Knowledge-Based Industries in India

M. Bandyopadhyay

Senior Expert, Centre for Science and Technology of the Non-aligned and other Developing Countries (NAM S&T Centre)

(Formerly, Adviser, International Cooperation, Department of Science and Technology, Government of India, New Delhi)

E-mail: bandodst@google.com

ABSTRACT

During the last six decades after its independence, India has achieved impressive success in the creation of S&T infrastructure and human resources, promotion of research and development, and the development, transfer and adaptation of technologies for industrial applications. Indian capabilities in science and technology cover a diverse range of disciplines, such as, advanced materials, agriculture, astronomy and astrophysics, biotechnology, chemicals and pharmaceuticals, information and communication technology, nuclear energy, oceanography, space technology and applications, etc. Changes have been carried out in the S&T strategy, policy and planning from time to time due to changing developmental needs of the country. Various mechanisms have been adopted and a number of schemes and programmes are being implemented by different S&T Ministries, departments and agencies for the development of an innovation ecosystem and promotion and development of knowledge-based industries. The important programmes in this regard include creation of institution-industry linkages, establishment of technology and business incubation facilities; providing support to innovators and S&T based entrepreneurship development; development and commercialization of technologies in knowledge-based areas; and financial assistance and fiscal incentives for the promotion of technology-based industries. This multi-pronged approach has derived favourable results with an impressive success in the knowledge-based industries and other skill oriented sectors which have been driving the economic progress in the country.

Keywords: *Entrepreneurship, Incubation, Industry, Innovation, IPR, Knowledge, Policy, R&D, S&T, University.*

Introduction

India has had a long and distinguished tradition in Science and Technology (S&T) which dates back to about 2500 BC. Innovative ideas relating to technology have been part of Indian culture and the basis of its civilization in the past. According to Archeological evidences, people in India had skills in the areas of agriculture; astronomy and mathematics; medicine and surgery; metallurgy; town planning and architecture. About 200 sites of Indus Valley Civilization are presently known which exhibit the use of scientific materials and innovative practices. More than 10,000 books were written in India on science and technology during the 12th to 18th Centuries (APCTT).

However, with subsequent historical and socio-political changes, there was a considerable setback and India could not keep pace with the revolutionary developments in science and technology in the western countries. Nevertheless, during the early Twentieth Century, the great discoveries and general scientific temper in the West largely influenced the Indian scientists, and the resurgence of Indian science started at that time. Some British scientists (*e.g.* Ronald Ross and others) working in India made brilliant contributions, and Indian scientists like C.V. Raman, J. C. Bose, P.C. Ray, Meghnad Saha, S.N. Bose, S. Ramanujan and others received wide international recognition for their outstanding work in various branches of science. This provided a multiplier effect on scientific activity and a favourable atmosphere was created for building a sound S&T base in the country.

Since independence, a large infrastructure has been built up with excellent facilities in various areas of S&T. Autonomous research councils were set up to coordinate and fund research in different sectors. Concomitantly, an administrative support structure was created to plan, evolve and administer various S&T schemes and programmes in the country. Scientific Ministries and departments, and S&T agencies in the public sector have established a large number of research and development (R&D) institutions. The higher education sector was strengthened by establishing more and more universities and institutions of national importance and also by introducing academic courses in modern disciplines of science, technology and engineering.

The Government has carried out appropriate changes from time to time in its approach and policies on S&T with a view to encouraging research and innovation that are relevant to the economy and society, promoting close and productive interaction between private and public S&T institutions, establishing an Intellectual Property Rights (IPR) regime, and creating a comprehensive national system of innovation covering science and technology, legal, financial and other related aspects.

This paper gives an overview of the S&T management structure and systems that have been developed in India over the years; and analyzes how the strategies and policies on science and technology have evolved towards developing an innovation ecosystem in the country. Finally, various mechanisms adopted and initiatives taken to promote and develop knowledge-based industries in the country have been reviewed.

Science and Technology in India

Science and Technology Management

In 1940s, a Board of Scientific and Industrial Research was constituted followed by the establishment of the Council of Scientific and Industrial Research (CSIR) in 1942. Soon after the independence, a strong link was established between politicians, decision makers and scientists, and Science and Technology was identified as a priority area for the country. Gradually, a large R&D infrastructure was built up with excellent facilities and research councils set up to coordinate and fund research in different sectors. During the First Five Year Plan (1951-56) itself, CSIR had established national laboratories for Physics, Chemistry, Metallurgy, Fuel, Glass and Ceramics, Food Technology, Drugs, Electrochemistry, Road research, Leather, and Building research.

A chronological account of the development of the R&D management structure in India is given in Table 8.1.

Table 8.1: Development of R&D Management Structure in India

Sl.No.	*Organization*	*Year Estd.*	*Objectives and Functions*
1.	Council of Scientific and Industrial Research	1942	Promotion, coordination, guidance and funding of scientific and industrial research
2.	Indian Council of Agricultural Research	1947	Coordination and funding of research, education and extension in agriculture, animal husbandry and fishery sciences
3.	Indian Council of Medical Research	1949	Coordination and funding of medical research
4.	Atomic Energy Establishment/Bhabha Atomic Research Centre	1954/ 1967	R&D on nuclear reactor design and installation, fuel fabrication, chemical processing of depleted fuel and development of radioisotope application techniques in medicine, agriculture and industries
5.	University Grants Commission	1956	Promotion and coordination of university education and research
6.	Defence R&D Organization	1958	To conduct defence related applied research and to support extramural research
7.	Indian Space Research Organization	1969	To develop space technology and its application to various national needs
8.	National Science and Engg. Research Board	2010	Funding of basic research in Science and Engineering with autonomy and flexibility

Soon after the independence, a Department of Natural Resources and Scientific Research was established in 1948, but later, an elaborate administrative support structure was created to plan, evolve and administer various S&T schemes and programmes in the country. The Department of Science and Technology (DST) which was established in 1971 is given the mandate to assist in formulation of policy statements and guidelines for S&T, to implement and co-ordinate activities which are multidisciplinary and involve several agencies and ministries, and to act as the Secretariat to apex advisory bodies on S&T. A number of scientific Ministries/ Departments were set up to coordinate and implement policies and programmes in specific areas such as, Department of Atomic Energy (1954); Department of Electronics (1970) [later renamed as Department of Information Technology which along with Department of Telecommunication formed the Ministry of Communication and Information Technology]; Department of Space (1972); Department of Agriculture Research and Education (1973); Department of Environment (1980) [later became Ministry of Environment and Forests]; Department of Defence Research and Development (1980); Department of Ocean Development (1981) [later clubbed with India Meteorological Department to become Ministry of Earth Sciences]; Department of Non-conventional Energy Sources (1982) [now named as Ministry of New and Renewable Energy]; Department of Scientific and Industrial Research (1985); Department of Biotechnology (1986), and Department of Health Research (2007). Apart from these, various socio-economic Ministries operate S&T schemes and programmes in their respective sectors.

S&T Systems

The S&T systems in the country consist of five main components *viz.* (*i*) Central (federal) government, (*ii*) State governments, (*iii*) higher education sector, (*iv*) public sector industry, and (*v*) private sector industry and non-profit institutions/scientific and industrial research organisations (SIRO). Various scientific Ministries and departments and S&T agencies under the Central Government have established a large number of R&D institutions. The State governments have also set up need based institutions for R&D and other scientific activities.

The higher education sector was strengthened by establishing a number of universities and institutions of national importance and also by introducing academic courses in modern disciplines of science, technology and engineering. As against only 20 universities and 500 colleges at the time of independence, currently there are 504 universities, including deemed to be universities and institutes of national importance, which are engaged in imparting higher education and conducting research in frontier areas of science and technology. As of December 2009, there were 25,951 colleges for higher education and 7272 degree level engineering and technology institutes in the country, including 65 centrally funded institutions (MHRD 2010). Thus a large contingent of S&T manpower of high caliber has been formed over the years, which was estimated to be 8.087 million in the 2001 Census out of which 391,000 were employed in R&D establishments (DST 2009). It has recently been decided to set up 30 new universities in the Central sector across the country, including 14 world class innovation universities in order to improve the number of high quality

S&T manpower. The government efforts on higher education and R&D have also been supplemented by gradual initiation of R&D activities by a number of public and private industrial enterprises and non-profit institutions/Scientific and Industrial Research Organisations (SIROs). A break-up of S&T institutions under various sectors in the country (up to 2009) is given in Table 8.2. These institutions are of varying size in terms of expenditure, manpower and infrastructure, besides variation in the quality of research performed by them.

Table 8.2: No. of S&T Institutions Under Various Sectors

No.	*Sector*	*No. of S&T Institutions (up to 2009)*
A1	Major S&T agencies of Central Government	462
A2	Central Government Ministries/Departments	245
A3	Public Sector in-house R&D institutes under Ministries/ departments	115
A.	**Total under Central Government**	**822**
B.	State Governments (Institutes, Research Stations, Joint Sector Companies)	834
C1	Central Universities	40
C2	State Universities	243
C3	Private Universities	53
C4	Deemed Universities	130
C5	Institutes of National Importance	38
C.	**Total Higher Academic Institutions**	**504**
D.	**Private Sector in-house R&D institutions and SIROs**	**2020**
	GRAND TOTAL	**4180**

Source: DST 2009, MHRD 2010.

S&T Planning and Funding

The Planning Commission was set up in 1950 which is responsible for the formulation of the five-yearly development plans of the country. The Commission decides the S&T plans and allocation of funds for various S&T programmes as a part of the overall socio-economic development plans. The emphasis laid by the government on S&T is reflected on the steady increase in the plan expenditure on S&T which rose from only ‘200 million in the First Plan (1951 – 56) to ‘733 billion in the Eleventh Plan (2007-12). It may be noted that the S&T allocation in the 5th five-year plan (1974-79) witnessed a four-fold increase over the 4th five-year plan (1969-74) as during this period the country's first S&T plan prepared by the NCST was integrated as a component of the socio-economic plan. Again, the outlay of six major scientific agencies has tripled from ‘253 billion in the Tenth Plan (2002-07) to ‘733 billion in the Eleventh Plan (Planning Commission, DST 2009).

The Gross Expenditure on Research and Development (GERD) of India has been less than one per cent of GDP so far, which is much less as compared with the developed countries. In the year 2005-06, the national investment on R&D was '287.78 billion, which was 0.88 per cent of GDP. The major portion (74.1 per cent) of the investment on R&D came from the government exchequer and the share of industrial sector towards the same has been only 25.9 per cent, which is 0.55 per cent of their sales turn-over (DST 2009).

Evolution of Science and Technology Policy

Advisory Mechanisms for S&T Policy and Planning

In order to identify priority areas of research and advise the government on S&T, several expert bodies and high level committees were constituted from time to time. The first such advisory committee was set up in 1948 under the Chairmanship of the first Prime Minister, Pandit Jawaharlal Nehru. Later, a Scientific Advisory Committee (SAC) was formed in 1958 which subsequently took various names and forms after several reconstitutions and with broader terms of reference as shown in Table 8.3. These committees were actively involved in the process of formulation of S&T policies and planning and to monitor the implementation of the same. As per recommendations of these bodies various mechanisms and programmes were evolved for strengthening the S&T capabilities of the country in different areas.

Table 8.3: Establishment of S&T Policy Making Advisory Bodies

Sl.No.	*Name*	*Year Estd.*	*Objectives and Functions*
1.	Scientific Advisory Committee to the Cabinet	1958/81/97	Policy planning and advising Central Cabinet on S&T
2.	Cabinet Committee on S&T	1968/81	– do –
3.	National Committee on S&T (NCST)	1971	Advice on inter-departmental cooperation and R&D–socio-economic links
4.	Commission for Additional Sources of Energy (CASE)	1981	To give greater thrust for the renewable energy sector
5.	National Biotechnology Board (NBTB)	1982	To accelerate development in Biotechnology
6.	National S&T Entrepreneurship Development Board (NSTEDB)	1982	To create self-employment opportunities for S&T persons and promote technology based enterprises
7.	National Council of S&T Communication (NCSTC)	1982	Popularization of science and creation of a scientific temper in the country
8.	Technology Policy Implementation Committee	1983	To suggest measures to follow up on Technology Policy Statement
9.	Scientific Advisory Council to the PM	1986/2005	Advising the Prime Minister on S&T
10	Technology Information Forecasting and Assessment Council (TIFAC)	1987	Strengthening national capability in technology forecasting and assessment and to provide advice

Contd...

Table 8.3–*Contd...*

Sl.No.	*Name*	*Year Estd.*	*Objectives and Functions*
11	National S&T Council (NSTC)	1990	Policy and direction for the development and application of S&T
12.	Technology Development Board (TDB)	1996	Support for development and commercialization of technologies
13.	National Innovation Council	2010	To evolve an Indian model of innovation focusing on inclusive growth

Changes in S&T Policy

The Science and Technology policy and strategy have undergone several changes and orientations from time to time as necessitated due to changing developmental needs of the country. By and large, Indian S&T policy has evolved in the last six decades after the independence in the following phases in tune with the changes in the country's industrial and economic policies (Bandyopadhyay 2011).

Formative Phase (up to 1960) – Development of S&T Infrastructure

After the independence, the Government of India identified Science and Technology as a priority area for the country. One of the important landmarks in the history of S&T in India was the adoption of the Scientific Policy Resolution (SPR) in 1958 which emphasised the role of S&T in the development and prosperity of the nation. The aims of the SPR were to foster, promote and sustain the cultivation of science and scientific research; to ensure an adequate supply of research scientists and to recognize their work; to encourage and initiate programmes for the training of S&T personnel; and to secure the people of the country all the benefits from the acquisition and application of scientific knowledge (GOI 1958). During the period up to 1960, the major thrust of the First (1951-56) and Second (1956-61) Five Year Plans was to build up a sound infrastructure on S&T with a chain of research laboratories and institutions in different parts of the country and create adequate S&T manpower (Planning Commission). Gradually, a large number of national research laboratories was set up with excellent facilities in various fields. Research councils such as Council of Scientific and Industrial Research (CSIR), Indian Council of Agricultural Research (ICAR) and Indian Council of Medical Research (ICMR) were created to coordinate and fund research in specific areas.

Protectionist Phase (1960 to 1980) – Emphasis on Technological Self-Reliance

During the early 1960s, several socio-economic developments led the Government to reorient its S&T strategies along with the economic policies. It was realized that unhindered import of technologies was not only leading the country to become dependent on the technology providers even at the cost of unfavourable conditions laid by them, but the sound R&D facilities that had already been established in the country were not being utilized for productive purposes. Therefore, the policy changes were brought about during this period to restrict monopolies, foreign investment and imports of technology. Efforts were made to find substitutes for imported goods and technology and thus a technology import substitution strategy was adopted. Further, in 1970, a Patent Law was enacted to safeguard the rights for process technologies, though it restricted product patenting.

During the period, technology import policy was made more restrictive and selective. Import of technology was allowed only in selected sophisticated and high priority areas where Indian skills and technology were still not adequately developed, with the condition that the importers would put in place suitable R&D efforts for adaptation and assimilation of such imported technologies for which several fiscal incentives were provided. The S&T policy focus of the Fourth Five Year Plan (1967-72) was on integrating industrial research, improving coordination and evaluation of research programmes periodically to achieve self reliance. For the implementation of these strategies, premier research institutions were directed to give priority to applied research aimed at technology import substitution which could lead to savings in foreign exchange.

Recognizing the role of technology, the S&T capability available in the country was sought to be exploited for industrial application. The development and application of appropriate technology, and development and widespread application of small and simple machines and devices for improving the productivity and earning capacity of workers in small and village industries were emphasised, and utilisation of indigenous technology for industrial application was advocated (GOI 1977).

In the 1970s, further S&T policy reorientation took place with the creation of National Committee on S&T (NCST) to facilitate inter-departmental cooperation and link scientific research with socio-economic and industrial needs. Department of Science and Technology (DST) was set up to coordinate implementation of the S&T policies and plans. For the first time, a separate S&T Plan was prepared as a component of the Fifth Five Year Plan (1974-79), in which technological capability was sought to be developed to meet the requirements of several identified sectors of economy. Emphasis was laid on the development of products and process technologies, and strengthening and utilization of existing S&T infrastructural facilities. A technical development fund scheme was initiated in 1976 to regulate and approve foreign exchange required for technology import including import of technical know-how, design and drawings, consultancy services, and plant and machinery.

As a result of these policy changes, R&D in India matured considerably into the development of industrially applicable products and processes in many areas such as agriculture and food processing, chemicals, electronics, drugs and pharmaceuticals, pesticides, etc. Significant technological capability was developed in several sectors such as heavy machines, electrical industries, mining machinery, railway stock, road transportation, and imports in many sectors was considerably reduced (GOI 1970). This also provided a favourable climate for the R&D institutions to create links with the industry and develop industrially applicable technologies, though in practice, not much progress was achieved in this regard at least up to 1980s.

Transition to Liberalization (1980 to 1990) – Towards Modernization and Technological Upgrading of Industry

It was suggested in the Seventh Five Year Plan (1985-90) that there was a need to consolidate on the strengths gathered so far and to take fresh initiatives to enable Indian industry to respond effectively to the emerging challenges. A beginning was made to reverse the restrictive policy directions of the 1970s and the reforms started during the 1980s gradually reduced the industrial licensing requirements, removed restrictions on investment and expansion, and facilitated easy access to foreign technology and foreign direct investment.

During this period, a number of policy and procedural changes were introduced aimed at facilitating transfer of technology, technological upgradation and modernisation of industry for increasing productivity and quality, improving competitiveness in the international market, and to reduce cost of production. Substantial resources for R&D were proposed to be allocated to constantly update technologies. Companies with in-house R&D establishments, and capability to absorb, adapt and disseminate modern technology were permitted to import technologies to increase their efficiency and cost-effectiveness.

While the basic ideology of harnessing the benefits of S&T was outlined in the Scientific Policy Resolution (SPR), the government of India felt the need to stipulate some policy guidelines for implementing the objectives of the SPR in the changed economic environment. As such, a Technology Policy Statement (TPS) was adopted in the year 1983 which highlighted the need for evolving suitable systems for the choice of technology, indigenous development and support to technology, and efficient absorption and adaptation of imported technologies in order to attain technological competence and self reliance and provide maximum gainful and satisfying employment to all strata of society (DST 1983). A Technology Policy Implementation Committee (TPIC) was constituted in 1983 to evolve necessary schemes and programmes to attain the goals set out in the TPS.

Subsequently, on the recommendation of the TPIC, a Technology Information Forecasting and Assessment Council (TIFAC) was established as an autonomous body for creating technology data bases and preparing state-of-the-art reports in various technology areas for use as essential tools for strengthening national capability in technology forecasting and assessment and for providing advice to the government on S&T planning. Further, in order to give greater thrust on new and emerging technologies, separate administrative departments were created in the Government in various sectors of S&T such as non-conventional energy, ocean development, environment and biotechnology.

A large number of schemes and programmes were also started for the promotion, development and transfer of industrial technologies during the Sixth (1980-85) and Seventh (1985-90) Five Year Plans. Six Technology Mission projects were launched to address major social objectives in drinking water, edible oils, tele-connectivity to all villages, dairy development, immunization and literacy. Another important initiative was to establish a National Science and Technology Entrepreneurship Development Board (NSTEDB) in 1982 to create a climate of entrepreneurship in higher educational institutions, generate self-employment opportunities for S&T qualified persons and promote knowledge driven and technology intensive enterprises in the country. Perhaps this was the major initiative of the Government for the first time to create linkages between the S&T institutions and industry through various programmes of the NSTEDB.

Separate policy initiatives were also taken during this phase in the information and communication technology sectors, which include, (*i*) Policy on electronics components – 1981, (*ii*) Industrial and licensing policy for colour television receivers – 1983, (*iii*) Measures to accelerate the rapid development of electronics – 1983, (*iv*) New Computer policy – 1984 and 1988, (*v*) Integrated policy measures in electronics – 1985, (*vi*) New Software Policy–1986 and (*vii*) World Market Policy – 1988.

Post-liberalization Phase (1990 onwards)

The Government decided to liberalize and open up the economy of the country in 1991, and a new Industrial Policy Statement (IPS) was announced in the same year, under which the distortions of the pre-liberalization period were corrected by dispensing with industrial licensing and facilitating foreign investment and technology imports, and at the same time taking steps to strengthen indigenous technological capability. The objectives of the IPS 1991 include encouraging entrepreneurship, development of indigenous technology through investment in R&D, transfer of new technologies, dismantling of the regulatory system, development of the capital markets and increased competitiveness for the benefit of common man (GOI 1991). Further, India ratified the agreement establishing the WTO, which came into force from January 1995, and as per the commitments under the TRIPS Agreement, the Patent Act was relaxed to allow product patents, which had positive impact on sectors like food, agro and pharmaceuticals.

Recognizing the need for developing a knowledge-based economy under the changing climate in the new era of globalization, the Government adopted a new Science and Technology Policy in 2003 (DST 2003). The primary aim of this policy includes encouraging research and innovation in the areas of relevance for the economy and society, particularly by promoting close and productive interaction between private and public S&T institutions, establishing an intellectual property rights regime, etc.

Further, in 2007, a national biotechnology development strategy was approved with the aim of laying a strong foundation for discovery and innovation, effectively utilizing novel technology platforms. A national biotechnology regulatory authority was also announced to create an independent, autonomous and professionally led body to provide a single window mechanism for biosafety clearance of genetically modified products and processes.

Developing an Innovation Ecosystem

India: An Emerging Power for Innovation

With the sound S&T infrastructure and highly skilled manpower, India has now become an emerging power for innovation. India has the following strengths that provide a very high potential for the country to grow as a knowledge economy (APCTT):

- Broad and diversified S&T infrastructure
- A strong manufacturing base
- Institutions for higher education
- A large technical manpower of high quality
- Valuable knowledge linkages and networks
- Global provider of software services
- Dynamic private sector
- Micro-economic stability
- Impressive Diaspora
- English speaking knowledge worker

The Science Advisory Council to the Prime Minister, in its vision document titled "India as a Global Leader in Science" (SAC-PM, 2010) opines: "*the potential for innovation is extremely high since diversity and adversity, both crucial to innovation, are characteristic features of the Indian scene. The first permits and encourages out-of-box thinking, while the latter throws up the day-to-day problems that can be overcome only by creative solutions*". The report recommends various measures to create an ecosystem to translate Indian potential into reality, that include, venture funding and tax incentives, encouragement to scientists and technologists to realize ideas into commercial ventures, extra-budgetary grants to government funded S&T institutions to cover risks on new ideas and innovation, encouraging tie-up among academia, R&D institutions and industry, strengthening IPR laws, and special package for promoting R&D intensive small firms.

Policy Initiatives for Innovation

The National Knowledge Commission (NKC) of India has identified innovation as one of the key factors in the economic growth of the country. NKC envisaged a national innovation system, where entrepreneurship at the local and national levels is encouraged, and inter-disciplinary studies in S&T are undertaken in order to encourage new approaches and methodologies (NKC). It was also decided in the S&T Policy 2003 to create a comprehensive national system of innovation covering science and technology and other related aspects. Accordingly, a series of policy measures have been proposed to facilitate the development of a strong and vibrant innovation climate in the country.

A draft National Innovation Act was proposed in 2008, which aims to facilitate public, private or public-private partnership initiatives for building an innovation support system, to encourage innovation, evolve a national integrated science and technology plan and codify and consolidate the law of confidentiality in aid of protecting confidential information, trade secrets and innovation. Subsequently, in pursuance of a recommendation of the NKC, the Government has proposed to enact a new legal framework, "Public Funded R&D projects (Protection, Utilization and

Regulation of Intellectual Property)" Bill (often referred to as India's Bayh-Dole Act), which is awaiting ratification by the Parliament. The bill aims to organize, promote and regulate the public availability of the intellectual property (IP) originating from government-aided R&D and for ownership and licensing of patents on outputs of public funded research.

The Government has declared 2010-2020 as the "Decade of Innovations". The main aim of this declaration is to develop an innovation ecosystem in the country to stimulate innovations and to produce solutions for the societal needs in terms of healthcare, energy, urban infrastructure, water and transportation. In pursuance of this declaration, a National Innovation Council (NIC) has been constituted to evolve an Indian model of innovation focusing on inclusive growth. The NIC has announced the setting up of a fund with a corpus of Rs 10 billion to encourage innovation. Further, a new Science, Technology and Innovation Policy has been drafted for wider consultations among various stakeholders, which is expected to be adopted in the near future.

The knowledge initiatives on innovation for the Eleventh Plan (2007-12) as recommended by NKC are (NKC):

- To put in place a National Innovation Policy which encourages competition among enterprises, greater diffusion of knowledge and increased support to early stage technology development initiatives and grass root level innovators.
- To foster increased collaboration among the R&D institutes, universities and private sector enterprises and leverage upon their cumulative strengths in designing and implementing various innovation programs.
- To create new interface structures to forge partnerships between academia and industry.

The focus of the S&T component of the Eleventh Five Year Plan (2007-12) includes enlarging the pool of S&T manpower and strengthening the S&T infrastructure to push India into the knowledge era as a global player and raising the Indian economy to the level of developed nations; emphasizing on knowledge based technologies *e.g.* information technology, nano-technology, biotechnology, and convergence of these technologies to make India a 'Global innovation leader' through a strong and vibrant innovation ecosystem (Planning Commission).

Some important measures were proposed in the Eleventh Five Year Plan (2007-12) for the introduction of technological innovation in small and medium enterprises (SMEs) which include, setting up a large number of technology business incubators and technology innovation centres; promoting academia-SME interaction for innovation; entrepreneurship development training, using polytechnics and industrial training institutes for producing multi-skilled human resources with the involvement of industry; and a mission mode approach for technology intervention in SME clusters (GOI, 2006).

Promotion and Development of Knowledge-Based Industries

Various mechanisms adopted, and schemes and programmes initiated by different S&T Ministries, departments and agencies for promoting innovations and developing knowledge-based industries may be by and large grouped under four major categories, *viz.* institution-industry linkages and technology business incubation; development and commercialization of technologies in knowledge-based areas; promotion of innovation and entrepreneurship; and financial assistance and fiscal incentives.

Institution-Industry Linkages and Technology Business Incubation

Development of institution-industry links is widely recognized to be of increasing significance for the creation of knowledge-based industries, and in the overall technological and industrial competitiveness of enterprises. Various types of linkages that have been found successful can be classified as, (*i*) Institution-based linkages *e.g.* industrial liaison office, technology transfer office, research consortia, industry cooperative, technology brokerage company; etc. (*ii*) Linkages through the creation of additional organizational infrastructure *e.g.* cooperative research centre, centre of excellence, incubation centre; etc. and (*iii*) large property based initiatives in/around institutions *e.g.* science park, innovation centre, technology park, research park, industrial park, technopolis etc. (Bandyopadhyay, 2008).

In India, the National Science and Technology Entrepreneurship Development Board (NSTEDB) implements schemes on Science and Technology Entrepreneurs Parks (STEPs) and Technology Business Incubators (TBIs), which are primarily aimed at creating institution-industry linkages and providing technological services and business incubation facilities for the promotion of knowledge driven and technology intensive enterprises. To encourage innovation and entrepreneurship, the Government of India has announced service tax exemptions with certain conditions for entrepreneurs physically located in a STEP or TBI recognized by NSTEDB.

Science and Technology Entrepreneurs Parks (STEPs) are established in and around S&T institutions for sharing ideas, knowledge, experience and infrastructural facilities, services and equipments of the institutions for the development of new technologies and their transfer to the entrepreneurs. A STEP provides nursery sheds, and offers technical support services such as quality assurance, testing and calibration facilities, precision tool room/central workshop, prototype development, consultancy, training, and business facilitation services such as computing, data bank, library and documentation, and common utility services for communication, office equipment, seminar hall/conference room, etc. So far, 15 STEPs are operational in different parts of the country, which have promoted nearly 910 units generating annual turnover of around ₹1.30 billion and employment for 6300 persons. More than 100 new products and technologies have been developed by the STEPs/STEP promoted entrepreneurs (ISBA 2009).

Technology Business Incubators (TBIs) promoted by NSTEDB are usually located in or around S&T institutions, or have strong links with such institutions. The TBI is an important mechanism for idea generation, commercialisation of research outputs of institutions, facilitating transfer of technology, and creation of new technology based and knowledge driven enterprises. Various services provided by a TBI to the entrepreneurs include work stations for a limited period; information on product ideas/technologies; business planning and training; common facilities for communication, conference, computers; facilities of the host institute at nominal charges; market survey/marketing assistance; technical assistance; assistance in obtaining statutory approvals, assistance to obtain finances; arranging legal and IPR services etc. Nearly 50 TBIs have been established in different parts of the country (ISBA 2009, DST 2010).

Development and Commercialization of Technologies in Knowledge-Based Areas

New Millennium Indian Technology Leadership Initiative (NMITLI)

The New Millennium Indian Technology Leadership Initiative (NMITLI) scheme is being implemented by the Council of Scientific and Industrial Research (CSIR), which is the largest public-private partnership effort to catalyse innovation centred technological development in selected areas by synergising the competencies of publicly funded R&D institutions, academia and industry. The programme has evolved 57 projects involving 80 industry partners, 270 R&D groups and 1700 researchers with an estimated outlay of '5 billion (CSIR). The projects are setting new global paradigms in the areas such as biotechnology, bio-informatics, drugs and pharmaceuticals, chemicals, materials, information and communication technology, weather forecasting, leather, energy etc.

Biotechnology

Department of Biotechnology (DBT) has launched Small Business Innovation Research Initiative (SBIRI) to bring biotechnology industry at the forefront of the technological revolution and provide financial support to enterprises to help early phase of product development. DBT has also started the Biotechnology Industry Partnership Programme (BIPP), to give a fillip to innovation and R&D in Biotechnology by promoting government partnership with industry. BIPP promotes high risk, transformational technology/process development and is directed at path breaking research in frontier futuristic technology areas. The emphasis of the programme is on health, agriculture, energy, and environment friendly manufacturing areas (DBT, 2010). Further, DBT has been setting up Biotech parks and incubators to facilitate innovation through the development of biotechnology industry clusters and help start-up companies and new entrepreneurs with the facilities for development and design of biotechnology products and processes.

Drugs and Pharmaceuticals

The Department of Science and Technology (DST) started a Drugs and Pharmaceuticals R&D programme in 1994-95 for promoting collaborative research in drugs and pharmaceuticals sector with the objectives of synergizing the strengths of public funded R&D institutions and Indian pharmaceutical industry; to create an enabling infrastructure, mechanisms and linkages to facilitate new drug development., and to enhance self-reliance in drugs and pharmaceuticals especially in areas critical to national health requirements. In 2004, the Government established Drug Development Promotion Board (DDPB) and set up a Pharmaceuticals R&D Support Fund (PRDSF) of ₹1500 million for supporting R&D projects jointly proposed by industry and scientists. So far, under this programme, 70 projects have been supported, 6 products developed and 12 process patents filed. Nine national facilities have been set up (DST 2010).

Another important innovation initiative in this area is the Open Source Drug Discovery (OSDD) project, which is a CSIR-led team India consortium with a global partnership, launched in September 2008 with a view to provide affordable healthcare to the developing world by providing a global platform where the best minds can collaborate and collectively endeavour to solve the complex problems associated with discovering novel therapies for neglected tropical diseases like malaria, tuberculosis, leishmaniasis, etc. The project provides a unique opportunity for scientists, doctors, technocrats, students and others with diverse expertise to work for a common cause. The Government of India has committed ₹1.50 billion (~US $30 million) towards this project. An equivalent amount of funding would be raised from international agencies and philanthropists (OSDD).

Renewable Energy

The Ministry of New and Renewable Energy (MNRE) is supporting application-oriented non-conventional energy technologies throughout the country. Research and development efforts of MNRE are directed to the improvement of efficiency, cost reduction, reliability, and durability of the new and renewable energy devices and systems. To encourage industry, Special Incentive Package Schemes have been announced that offer various incentives and capital subsidies for investment on setting up of solar PV manufacturing facilities. The Indian Renewable Energy Development Agency (IREDA) and renewable energy development agencies at various states are undertaking several programmes and providing subsidies for producing and using non-conventional energy devices. A National Solar Mission has been launched with the aim to set up an enabling environment for solar energy penetration in the country with a target to develop and deploy 20,000 MW solar power, including 20 million solar lighting systems in rural areas by 2022 (MNRE 2010).

Information and Communication Technology (ICT)

As a part of the reforms agenda, the Indian Government has taken major steps to promote ICT, with a focus on software development for export; telecommunications policy reform, privatisation of the national long-distance and mobile phone markets; and development of a more comprehensive approach to ICT.

Software Technology Parks of India (STPI) was set up by the Department of Information Technology (DIT), Government of India in 1991 in order to promote the development and export of software and software services including Information Technology enabled services (ITeS) and Bio-IT; to provide statutory and other promotional services to the exporters by establishing Software Technology Parks (STP) and Electronics and Hardware Technology Parks (EHTP) and other such schemes; to provide data communication services including value added services to industries. STPs provides consulting, training and implementation services including network design, system integration, installation, operations and maintenance of application networks and facilities in varied areas and incubation facility to cater to the needs of SMEs. Till 2008-09, total 8455 units were operational in 51 STP centres/ sub-centres located in different parts of the country (STPI, 2009). Further, under the scheme on innovation promotion in IT the DIT provides support to specified institutions to promote start up companies.

Nano-Science and Technology

Several Government departments and S&T agencies such as DST, DBT, CSIR, DIT, DAE, DRDO, ICMR and MNRE are supporting R&D in nano-science and technology. The Government of India, in May 2007, launched a Mission on Nano Science and Technology (Nano Mission) with an allocation of ₹10 billion for 5 years with the objective of developing products and processes, especially in areas of national relevance like safe drinking water, materials development, sensors development, drug delivery, etc. For this, it will forge linkages between educational and research institutions and industry and promote public-private partnerships. 130 projects have been supported to individual scientists so far.

Nano-technology research in India is almost entirely a Government supported initiative, and the participation of industry is negligible. However, though the projects are largely on basic research, there have been some efforts to develop industrially applicable products and processes through public-private partnerships particularly in health, life sciences, materials, water purification etc. According to a status report on nano-technology developments in India, 21 technologies have been transferred by R&D/academic institutions to industry (TERI 2009).

Promotion of Innovation and Entrepreneurship

Technopreneur Promotion Programme (TePP)

The Department of Scientific and Industrial Research (DSIR), Ministry of Science and Technology is implementing the Technopreneur Promotion Programme (TePP) in order to provide assistance to the individual innovators to become technology-based entrepreneurs (or Technopreneurs), and to assist the technopreneurs in networking and to forge linkages with other constituents of the innovation chain for commercialization of their ideas and inventions. Under TePP, financial support is provided to selected individual innovators having original ideas for converting them into working models, prototype development etc. So far, support has been provided to 375 projects under TePP and 41 projects under micro techno-entrepreneurship programme (DSIR).

S&T based Entrepreneurship Development

Besides the schemes on STEPs and TBIs, the NSTEDB has also been implementing several other activities on training and awareness on S&T based entrepreneurship development. Following two activities have direct bearing on the promotion of technology-based enterprises (NSTEDB):

- Organization of Entrepreneurship Development Programmes (EDPs) to train S&T qualified persons in setting up of industry. 1130 EDPs/have been organized and 25,000 S&T persons trained in entrepreneurship so far.
- Establishment of Entrepreneurship Development Cells (EDCs) and Innovation and Entrepreneurship Development Centres (IEDCs) in S&T academic institutions to create an entrepreneurial culture and to foster techno-entrepreneurship. More than 40 EDCs/IEDCs have been set up.

Innovative Clusters

The Department of Science and Technology (DST) has initiated a programme on Promoting Innovative Clusters (PIC) to encourage innovation among micro, small and medium enterprises (MSMEs). The project aims to promote collective research and development and foster learning to develop new products, processes, technologies and services in clusters by facilitating inter-firm and inter-institutional linkages. The project envisages the development of cluster based 'innovation ecosystem' at national, regional and local levels. The identified industrial sectors include knitwear machinery, medical devices; medical electronics; perfumery; forge and foundry; agricultural implements; and sports goods (PIC).

Grassroots Innovations

The National Innovation Foundation (NIF) was set up by DST as an autonomous organization with the objective of providing institutional support in scouting, spawning and sustaining grassroots innovations and helping their transition to self-supporting activities. The NIF has network partners, the Honeybee Network and Society for Research and Initiatives for Sustainable Technologies and Institutions (SRISTI) which have been scouting innovations by farmers, artisans, women, etc. at the grassroots level, and Gujarat Grassroots Innovations Augmentation Network (GIAN) that scales up innovations. So far, more than 100,000 innovations and traditional knowledge practices have been scouted from 500 districts of the country. The areas covered in the innovation database include plant variety, utilities and general machinery, farm implements, energy devices, agricultural and traditional knowledge practices, livestock management, herbal remedies, biodiversity examples, innovation concepts and ideas (NIF).

Financial Assistance and Fiscal Incentives to Technology-Based Industries

Support from Technology Development Board (TDB)

The Government of India created a fund for technology development and application out of the proceeds of the Research and Development Cess levied (@5 per cent) on the imports of technology. To administer the Fund, a Technology Development Board (TDB) was set up in 1996, as a statutory body to promote development and commercialization of indigenous technology and adaptation of imported technology for wider application. The Board provides financial assistance in the form of equity soft loans to industrial concerns at a simple interest rate of 5 per cent per annum, or grants to R&D institutions. Up to 2005-06, the Board has supported 156 projects with financial assistance of ₹5.60 billion. A few notable success stories are the projects of M/S Shantha Biotechnics and M/S Bharat Biotech International, both for the development and production of recombinant Hepatitis-B vaccine using innovative technologies, and Eicher Motors for the introduction of a four cylinder E483 engine with innovative design and concepts (TDB).

Fiscal Incentives for R&D by Industry

Several fiscal incentives such as duty free imports, weighted tax deductions, tax holiday and income tax exemption are provided by Department of Scientific and Industrial Research (DSIR) to companies for undertaking industrial R&D. With effect from the financial year 2010-11, weighted deduction is allowed @ 200 per cent of the expenditure incurred on scientific research and development by companies in an approved in-house R&D facility, and weighted deduction @ 175 per cent is available for contribution to approved entities for the purpose of scientific research. Other incentives to encourage R&D by industry include write-off of revenue and capital expenditure on R&D, weighted tax deduction on sponsored research programmes of industry with national laboratories/universities; accelerated depreciation allowance on plant and machinery, customs duty exemption on goods imported for use in Government funded R&D projects, and excise duty waiver for three years on goods produced based on indigenous technologies and duly patented (DSIR).

Venture Capital Funding (VCF)

There are several Venture capital fund (VCF) firms, private equity funds and angel investors that are operating in the country for providing financial assistance to start-up technology-based enterprises in form of equity participation. Currently, VCFs are concentrating on IT/ITeS, healthcare and biotechnology, telecom, food technology, entertainment, education, new materials etc. There have been about 150 active funds in last three years, and in 2007 a total investment of US$14 billion was reported to have been made through 387 deals.

Support from Banks

Development financing institutions such as the Small Industry Development Bank of India (SIDBI) and Commercial Banks have set up several funds to provide support to SMEs for modernisation and technology/quality upgradation. These initiatives include (*i*) Technology Development and Modernisation Fund (TDMF), (*ii*) National Venture Fund for Software and Information Technology (NFSIT), (*iii*) Credit Grant Fund Trust for Micro and Small Enterprises (CGTMSE), (*iv*) SME Growth Fund and (*v*) Microfinance Programme (MSME; SIDBI).

Summary, Observations and Conclusions

With the encouragement and support provided by the Government, a sound infrastructural base for science and technology has been created in the country. These include research laboratories, higher educational institutions and highly skilled human resource. Indian capabilities in S&T cover an impressive range of diverse disciplines, such as Agriculture; Astronomy and Astrophysics; Biotechnology, Chemicals; Defence research; Electronics and Information Technology, Healthcare,

Drugs and Pharmaceuticals; Materials; Nuclear energy, Oceanography; Space technology and applications, etc.

Policy Implications

The Scientific Policy Resolution 1958 and the Technology Policy Statement 1983 enunciated the principles on which the growth of science and technology in India has been based over the past several decades. These policies have emphasized self-reliance, and sustainable and equitable development. With the changes in the economic policy, S&T Policy directions were further modified in 2003 keeping in view the challenges of globalization and the need for developing a knowledge-based economy.

Although the National Expenditure on R&D remained less than one per cent of GDP so far, in absolute terms, the investment had an increase of more than 200 per cent in the last five years, from ₹180.88 billion in 2002-03 to ₹377.78 billion in 2007-08. In 2003, the Government reaffirmed its earlier commitment to increase the R&D expenditure to 2 per cent of GDP by 2007. However, this did not happen, and the same was only marginally increased from 0.80 per cent in 2002-03 to 0.88 per cent in 2007-08 (DST 2009). Nevertheless, this reflects an emphasis of the Government on R&D and innovation in the S&T policy and strategy. The immediate follow up of such strategy has been a huge increase of almost 300 per cent in the financial outlay for the six major S&T departments, *viz.* Department of Atomic Energy (DAE), Ministry of Earth Sciences (MoES), Department of Science and Technology (DST), Department of Biotechnology (DBT), Department of Scientific and Industrial Research (DSIR/CSIR) and Department of Space (DOS) to ₹753 billion in the Eleventh Five Year Plan (2007-12) from ₹253 billion in the Tenth Plan (2002-07) (Planning Commission, DST 2009).

There has been another positive trend with regard to the participation of the business sector in the Gross Expenditure on R&D (GERD), which has increased from 18 to 28 per cent of GERD during the period from 2003 to 2007, which reflects a dynamic and innovative private sector. Due to the favourable climate and promotional policies on R&D and innovation, a number of multi-national corporations have come forward to establish R&D centres in India, mostly in ICT, automobiles, biotechnology and drugs and pharmaceuticals sectors, and there has been a tremendous increase in the number of foreign company owned R&D centres from less than 100 in 2003 to about 750 in 2009 (UNESCO, 2010).

Successful Mechanisms and Programmes

As follow up of the policy initiatives of the Government from time to time, a host of measures were taken for developing an innovation climate and promotion of a knowledge based economy. Various mechanisms have been adopted and a number of schemes and programmes initiated to encourage linkages between public funded S&T institutions and industry, and for the promotion and development of technology intensive industries. The Government has also taken several measures for establishing an intellectual property rights regime, including amendments in patent laws, modernisation of patent offices, and awareness programmes on IPR and related issues.

Technology incubators and science parks have been found to be the successful mechanisms for promoting high tech development and technological innovation in various parts of the world. In many countries, incubation centres are located in academic or R&D institutions which facilitate interaction between the institutions and entrepreneurs and play an important role in innovation and transfer of technology. In a study carried out by the author in the University of Oxford on the British model of science parks and the role of academic and research institutions in introducing technological innovation in industry, it was observed that the first ever science park was established in the United Kingdom in 1972, and by 1990, the UK Government and universities had invested about £300 million in developing high technology incubation centres and science parks (Bandyopadhyay, 1995). Further, in a case study it was reported that the University of Warwick had established 52 different kinds of linkages with 886 companies including MNCs, and the Science Park in the University had played an important role to attract knowledge-based companies on site, with more than 350 per cent increase in number of on-site companies within a span of seven years (Bandyopadhyay, 1994).

In India, a concrete step on institution-industry linkages and S&T parks was taken only in the 1980s when programmes on S&T Entrepreneurs Park (STEP) and S&T based entrepreneurship development were conceptualized and initiated under the National Science and Technology Entrepreneurship Development Board (NSTEDB). The activities of NSTEDB have shown good results in creating an entrepreneurial culture in and around academic institutions, forming institution-industry linkages, and establishing knowledge driven and technology intensive small enterprises. An evaluation study conducted by National Council of Applied Economic Research (NCAER 1998) shows that NSTEDB programmes have resulted in a social yield of internal rate of return of around 30 per cent. It was further noted that though the gestation period was quite high, the STEPs established by NSTEDB are successfully working with significant results in terms of no. of start-up firms established, new technologies developed and employment generated. The initial results of the programmes on STEPs and TBIs are encouraging, and the leadership role in enterprise building in these initiatives is very prominent that ensures the success of an enterprise in the commercialization of technologies involving risks (Nagayya 2006).

The performance of the Technology Development Board (TDB) has demonstrated coordinated efforts supported by efficient management, with cooperation by all the stakeholders, *viz.* the government, the R&D institutions the industry. There have been several success stories on developing knowledge based industry with TDB support (TDB). Other important initiatives are the NMITLI scheme of CSIR and TePP of DSIR. NMITLI enjoys a good brand image and is looked upon as a role model as India's largest public-private partnership R&D programme, and the industrial partners have highly benefited through high quality technological inputs from their R&D partners (CSIR). Further, during the last ten years of its operation, TePP has been able to identify several innovative ideas, and support has been provided under the programme to the innovators to develop and nurture them into products and processes at the commercial level (DSIR).

Status of Knowledge-Based Industries

Continuing changes in S&T policy coupled with the industrial and developmental policies have fueled rapid growth in various sectors of industry during last two decades in terms of variety, quality and quantity. Impressive indigenous capabilities have now been established to the point of self-sufficiency in many sectors. With the help of a proactive policy and liberal fiscal incentives, knowledge-based industries in biotechnology, drugs and pharmaceuticals, information and communication technology, renewable energy etc. have brought a revolution in the economy of the country.

Biotechnology Industry

The Government has taken several steps to create excellent facilities for research and development and skilled manpower on biotechnology, as a result of which it is now one of the fastest growing knowledge-based sectors in India with a threefold growth in the last five years with revenue of US$ 3 billion in 2009-10. Further, as compared to other sectors of industry, the biotechnology companies including several multi-national corporations have made considerable investment in R&D, and the biotech industry in the country is now capable to face the global challenges and competition.

Drugs and Pharmaceuticals Industry

As per the commitments under the TRIPS Agreement, the Patent Act was revised in 1995 with the provision to allow product patents. Contrary to the apprehensions that this would affect the indigenous drugs and pharmaceuticals industry, the new provisions of the Act rather had a positive impact, and the industry has been growing at a much faster rate after 1995. Moreover, it has been noted that R&D investment by private pharmaceutical firms has increased @35 per cent per annum in the last ten years. The drugs and pharmaceuticals industry is now one of India's foremost technology-based sectors, with wide ranging capabilities in drug development and manufacturing technology. India is 3rd in the world in terms of volume of production by this sector with a 10 per cent share in the global market. There are about 5000 firms in the country engaged in the manufacture of drugs and pharmaceuticals, which provide direct employment to about 340,000 persons. According to an estimate, the Indian pharmaceuticals industry has grown from only US$ 0.300 billion in 1980 to about US$ 19 billion turnover in 2008 (UNSCO, 2010).

Information and Communication Technology (ICT) Industry

This is another knowledge-based industry that has become the fastest growing sector in the country in the fast changing world market. The Department of Information Technology (DIT) has taken several policy measures including fiscal incentives, enhanced credit flow, and infrastructure and incubation facilities, due to which the IT industry has received tremendous opportunities in ITeS and BPO, products and components, and in e-Governance, education, transportation, etc. The IT-ITeS industry's contribution to the national GDP is around 6 percent, and annual revenues from outsourcing operations in India alone was US$50 billion in 2008-09, which is expected to increase to more than US$200 billion by 2020. The production of electronics and IT-ITeS industry in 2008-09 was estimated to be ₹3724.5 billion (DIT, 2010).

Renewable Energy

In this sector, a total grid-connected renewable (wind, small hydropower, biomass and solar power) power generation capacity of 15,694 MW was achieved till the end of 2009, which is about 10 per cent of the total installed power generating capacity in the country. The renewable power generating capacity has been planned to be increased to nearly 25,000 MW by 2012. There are about 90 companies in Solar Photo-voltaic (SPV) areas, engaged in manufacture of solar cells (~15 units), and PV modules (20 units) and rest in assembly and supply of SPV systems. During 2007, 45 MW of solar cells and 80 MW of SPV modules were produced out of which over 60 MW of SPV products were exported. Further, India is the world's 5th largest in wind energy with an installed capacity of 11806 MW (MNRE 2010). There are at least ten major companies with substantial capacities and investment in the area. Suzlon has a major share in the global wind turbine market, and is currently the leading manufacturer in India which has made India the leader in amongst the countries in advanced wind turbine technology. There is a huge untapped potential in this field that provides a great scope for the industry.

Automotive Industry

It has made rapid strides in India after de-licensing and opening up of the sector in 1991. It has witnessed the entry of several new manufacturers with the state-of-art technology. The import of technology for technology upgradation is also allowed in this sector under certain terms. The Indian automotive industry has already attained an investment of US$ 13.5 billion and a turnover of ~US$ 35 billion. The industry provides employment to 500,000 persons directly and to another 10 million indirectly. A National Automotive Testing and R&D Infrastructure Project (NATRIP) has been launched to create core global competencies in automotive sector and develop state-of-the-art testing, validation and R& D infrastructure in the country with a view to support the growth and development effort of the automotive industry to reach international levels.

Conclusions

India is increasingly becoming a leading global innovation player. The country achieved an impressive growth in the recent years, even during the severe economic recession in other parts of the world, which is primarily due to progressive policies and innovative practices for developing capability in S&T and to encourage knowledge-intensive manufacturing and services.

Recognizing the importance of development and commercialization of technologies in knowledge-based areas, the government of India has provided appropriate policy and regulatory framework and initiated a number of promotional measures and support services, leading to the significant success in the development of industries in sectors like Biotechnology, Drugs and Pharmaceuticals, ICT, Automobiles etc.

While the government should maintain the trend in the increase in the public sector expenditure in R&D, further impetus should be given in form of incentives and benefits to industry for R&D investment. In order that India remains globally competitive, it is necessary to strengthen the on-going efforts to foster linkages between higher academic institutions and public funded research laboratories on the one hand and the industry on the other, and expand the programmes on Science and Technology Parks and Technology Business Incubators with liberal funding from government and government owned developmental financing institutions for establishing such facilities. Support should be provided for the commercialization of indigenously developed technologies through the mechanisms of NMITLI and TDB by involving more industrial partners. Research scientists in institutions must be suitably motivated to accept assignments from industry with rewards and appropriate share of technology transfer income, and by allowing them to acquire equities in companies. The proposed Public Funded R&D projects Bill should be put in action at the earliest in order to streamline the utilization of the intellectual property arising out of government-aided R&D for industrial use, which would ensure a quality improvement in the IPR regime of the country. Programmes on entrepreneurship development and skill formation training should be strengthened to cover commercially relevant knowledge, technology and skills and to improve the productivity of the vast workforce of the country.

References

APCTT. *National Innovation Systems: A Case of India*. Asia Pacific National Innovation Systems on-line Resource Centre, APCTT (www.nis.apctt.org).

Bandyopadhyay, M., 1994. Technology management in academic institutions for economic growth and regional development: The case of an entrepreneurial university in England. *MDI Management Journal*, Gurgaon, 7(1).

Bandyopadhyay, M., 1995, Promotion of technology based industry and regional economic development by utilizing the expertise of academic and research institutions in Britain: A role model for India. In: *Science, Technology and Industrial Development in India*, (Eds.) S.C. Pakrashi and G.P. Phodke. Wiley Eastern Ltd., New Delhi.

Bandyopadhyay, M., 2008, Collaborative links between academic and research institutions and industry for stimulating technological innovation and economic development: Need for science and technology policy initiatives and networking amongst NAM and other developing countries. In: *Science and Technology Policy and Indicators for Development: Perspectives from Developing Countries*, (Eds.) Seetha I. Wickremasinghe and V.K. Gupta. Daya Publishing House, Delhi.

Bandyopadhyay, M., 2011, Policies and programmes for technological capacity building of small and medium enterprises in India. In: *Technology Transfer and Small and Medium Enterprises in Developing Countries*, (Eds.) K. Ramanathan, Keith Jacobs and M. Bandyopadhyay. Daya Publishing House, Delhi.

CSIR, Council of Scientific and Industrial Research (CSIR), Government of India, New Delhi. Website:http://www.csir.res.in

DBT, 2010. *Annual Report 2009-10*. Department of Biotechnology, Government of India, New Delhi.

DIT, 2010. *Annual Report 2009-10*. Department of Information Technology, Government of India, New Delhi.

DSIR, Department of Scientific and Industrial Research, Ministry of Science and Technology, Government of India, New Delhi. Website: www.dsir.gov.in

DST, 1983. *Technology Policy Statement*. Department of Science and Technology, Government of India, New Delhi.

DST, 2003. *Science and Technology Policy*. Department of Science and Technology, Ministry of Science and Technology, Government of India, New Delhi.

DST, 2009. *R&D Statistics 2007-08*. Department of Science and Technology, Ministry of Science and Technology, Government of India, New Delhi.

DST, 2010. *Annual Report 2009-10*. Department of Science and Technology, Ministry of Science and Technology,Government of India, New Delhi.

GOI, 1958. *Scientific Policy Resolution*. Government of India.

GOI, 1970. *Economic Survey 1969-70*. Government of India.

GOI, 1977. *Industrial Policy Statement*. Ministry of Industry, Government of India.

GOI, 1991. *Statement of Industrial Policy*. Ministry of Industry, Government of India.

GOI, 2006. *Report of the Steering Committee on Science and Technology for the Eleventh Five Year Plan (2007-12)*. Office of the Principal Scientific Adviser to the Government of India, New Delhi.

ISBA, 2009. *First Status Report on Technology Business Incubation in India*. Indian STEP and Business Incubators Association (ISBA).

MHRD, 2010. *Annual Report 2009-10*. Ministry of Human Resource Development (MHRD), Government of India, New Delhi.

MNRE, 2010. *Annual Report 2009-10*. Ministry of New and Renewable Energy (MNRE), Government of India, New Delhi.

MSME, Ministry of Micro, Small and Medium Enterprises, Government of India, New Delhi. Website: www.msme.gov.in

NIF, National Innovation Foundation, Website: www.nif.org.in

Nagayya, D., 2006. Perspectives of entrepreneurship development, and innovation and business incubators. In: *Banking Sector and Human Resources: Changing Scenario,* (Ed.) T. Sreenivas. Discovery Publishing House, New Delhi.

NCAER, 1998. *Evaluation of NSTEDB Programmes and its Impact on Entrepreneurship Development, Employment Generation and Technology Transfer*. Study sponsored by Department of Science and Technology, Government of India, New Delhi.

NKC, National Knowledge Commission, Government of India, Website: www.knowledgecommission.gov.in

NSTEDB, National S&T Entrepreneurship Development Board, Department of Science and Technology, Government of India, New Delhi. Website: http://www.nstedb.com

OSDD, Open Source Drug Discovery Project, Website: www.osdd.net

PIC, Promoting Innovation Clusters in India (www.innovationclusters.org)

Planning Commission, Government of India, New Delhi. Website: www.planningcommission.gov.in)

SAC-PM, 2010. *India as a Global Leader in Science.* Science Advisory Council to the Prime Minister.

SIDBI, Small Industries Development Bank of India, Website: www.sidbi.com

STPI, 2009. *Annual Report 2008-09.* Software Technology Parks of India, Department of Information Technology, Government of India, New Delhi.

TDB, Technology Development Board, Government of India, New Delhi. Website: *www.tdb.gov.in*

TERI, 2009. *Nanotechnology Developments in India: A Status Report, Capability, Governance and Nanotechnology Developments–A Focus on India,* The Energy and Resources Institute, New Delhi (Project Report No. 2006ST21:D5).

UNESCO, 2010. *UNESCO Science Report 2010.* United Nations Educational, Scientific and Cultural Organization, Paris, France.

Chapter 9

Public Understanding of Science and Technology: A National Prototype for Popularization of Science, Policy and Actions in Islamic Republic of Iran

***Mohammad Hassanzadeh*[1] *& Azadeh Heidari*[2]**

[1]*Faculty Member, Tarbiat Modares University, Iran*
E-mail: hasanzadeh@modares.ac.ir
[2]*Ph.D Student of Library and Information Science,*
Islamic Azad University, Science and Research Branch, Iran
E-mail: Az.heidari@gmail.com

ABSTRACT

In recent years, the presentation of scientific knowledge to the general public has had a significant place in public understanding of science. Governments, especially in developed countries, have been trying to make science and technology an essential part of the culture of the people. In this regard, they strength science, technology and innovation along with popularization of science and technology in order to communicating and appropriating scientific and technological knowledge for broad sectors of the population. Statistical reports show that Islamic Republic of Iran has a relatively well developed science and technology infrastructure among some other developing countries. Science and technology in Iran are increasingly recognized as the determinant factor of national development. On the other hand, studying the history of science in Iran has revealed that science has acquired a high status in the Islamic and Iranian culture. Despite the fact that Iran has a bright history of science and civilization, but the quantity and quality of public cooperation with scientists, science innovators, science policy makers, science decision makers, science centers, administrators and etc., need more attention in order to gain proper Public Understanding of Science (PUS) and as a result, sustainable development, which can be strengthened through Popularization of science activities.

This paper aims to discuss about 1) concepts and the role of popularization of Science in increasing the scientific knowledge in the country and Public Understanding of Science (PUS), briefly, 2) introducing one of the main administrator and its department, which is dedicated to Popularization of Science in Iran, and ends with introducing a recent national prototype for Popularization of Science, policy and actions in Islamic Republic of Iran.

Keywords: *Science, Technology, Public Understanding of Science (PUS), Popularization of Science, Islamic Republic of Iran.*

Introduction

In recent years the presentation of scientific knowledge has had a significant place in countries. Nowadays, governments are seeking to bridge the science gap that exists between scientific community and people in general.

In the age of the early Europe, several enlightened philosophical aspects rose to view. Francis Bacon's idea about scientific knowledge as a kind of useful knowledge, Rene Descartes practical philosophy about beneficial science, or Robert Boyle's experimental philosophy are the most relevant representatives of that period from our point of view. As a result of these early modern thoughts, we can observe how scientist's knowledge could play an increasingly important role in society, and how it started to behave as an objective background or as a social good. (Redey)[Online]

On the other hand, in recent times, science and technology is an essential element of socio-economic and cultural development and it needs appropriate national S&T policies for more growth.

So, utilizing from appropriate popularization of science techniques and increasing the public understanding of science (PUS) and public awareness of science will prepare a good situation for increasing scientific knowledge in the community and sustainable development will occur.

Scientists, science centers, universities and other science administrators, try to increase the science production, but how many people, lay people, rural people, children and other groups make benefit from scientific knowledge in their life? How much they know about science and to what extent they make use of it? It seems that a national prototype is necessary to be proposed. Needless to say, it is clear that for best result, collaboration in national scale is inevitable.

Science and Technology

The importance of science and technology to societies, especially to modern societies, and the role of a public awareness of science in sustainable development have long been recognized. But what does the "Science" mean?

Science

Science is a powerful means of understanding the world, the world in which we live and it is also capable of yielding enormous returns that directly enhance socio-economic development and the quality of our lives (World Conference on Science, 1999).

Science is part of the cultural stock of knowledge with which everybody ought to be familiar (Bauer *et al.*, 2007).

In common usage, the word science is applied to a wide variety of disciplines or intellectual activities which have certain features in common. Application of the term did not begin with any formal definition; rather the various disciplines arose independently, each in response to some particular need (McGraw-Hill Encyclopedia of Science and Technology, 1960).

The American College Dictionary describes Science as a branch of knowledge or study dealing with a body of facts or truths systematically arranged and showing the operation of general laws; 2- systematic knowledge of the physical or mathematical world, 3- systematized knowledge in general, 4- knowledge, as of facts or principles; knowledge gained by systematic study, 5- a particular branch of knowledge and 6- skill: proficiency (The American College Dictionary, 1963).

So, public understanding of science activities through popularization of science seems inevitable and governors and policy makers should pay more attention to it, as much as they think and plan for science production and for S&T policy making.

Public Understanding of Science

In the second half of the 1980s, new concerns emerge under the title "public understanding of science"(PUS) (Bauer *et al.*, 2007).

PUS was also extended to PUST to include "T" for technology, PUSTE, to include "E" for engineering or PUSH, to include "H" for the humanities, the last indicating a more continental understanding of "science" as "wissenschaH". The dating of these phases follows mainly the influential UK experience. In the US all through the 1970s the AAAS[1] had a standing committee on public understanding of science (Kohlstedt *et al.*, 1999).

So, a strong program for the popularization of science needs to be located in similarly strong institutions that carry a clear mandate for this purpose.

Popularization of Science

Popularization of science is nothing else than an endeavour to image scientific ideas in such a way that everyone (especially non-scientists) can grasp the fundamental concepts and have an idea of what science in essence is. Although nobody can tell exactly what "science" is all about, everyone should have an idea

---

1 The American Association for the Advancement of Sciences, triple A-S (AAAS), is an international non-profit organization dedicated to advancing science around the world by serving as an educator, leader, spokesperson and professional association (AAAS).

anyway. The question here is whether this is possible and, if so, to what extent (Cornelis).

The popularization of science and technology plays an important role in the socio-economic, cultural and environmental development of the countries. It fosters creativity and innovation and contributes to producing better trained human resources, expands social opportunities, and strengthens the educational system. Culturally and environmentally, the popularization of science and technology enhances the critical sense of the population, by increasing its involvement in decision-making and contributing to democratic stability and sustainable development. It helps to enhance personal satisfaction and self-esteem in the population. At present, with the growing importance that science and technology have taken on in all areas of social life, the popularization of science is increasingly becoming a very significant strategic issue [The workshop on the popularization of S&T (a report), 2004].

The rapid growth and unprecedented influence of new technologies, especially the information and communication technologies including the internet, is raising global awareness of the power of technology as a whole. The key issue should be the empowerment of individuals and groups to be able to use scientific knowledge and technological know-how to address such pressing problems as prevention public health crises, and other problems.

Initiatives in public understanding and utilization of science would consist of finding the means to deliver scientific information to lay people in a manner that makes it attractive and enjoyable.

The challenge to develop strategies that promote a science-technology culture has taken on greater significance because the popularization of science and technology takes place in a wide variety of practices. It also generally seeks to:

- ☆ encourage society at large to share in the fruits of science,
- ☆ make careers in scientific fields more attractive, accessible and beneficial to a wider sector of society, especially women,
- ☆ demystify scientific phenomenon through the application of science and technology to everyday problems,
- ☆ sensitize the population at every level to the daily occurrences and applications of science,
- ☆ promote a view of science as "fun" and enjoyable, especially to encourage active participation by children, and
- ☆ use science and technology in addressing the present crises that are the daily concerns of the population [Making science and technology work for the poor and for sustainable development in Africa: parliamentary documentation, Sustainable development (CSD-3), 2003].

At the Budapest world conference on science, "science for the twenty-first century; a new commitment" on 26 June-1 July 1999, some issues about science, technology and scientific knowledge were discussed. In this context, participants mentioned that: "... the sciences should be at the service of humanity as a whole and should contribute to providing everyone with a deeper understanding of nature and society, a better quality of life and sustainable and healthy environment for present and future generations ..." this commitment has several key points and considering points. One of its proclaims is stressing on science education and popularization of science. "Science education, in the broad sense, without discrimination and encompassing all levels and modalities, is a fundamental prerequisite for democracy and for ensuring sustainable development. It is on this platform that science education, communication and popularization need to be built. It is more than necessary to develop and expand science literacy in all cultures and all sectors of society as well as reasoning ability and skills and an appreciation of ethical values, so as to improve public participation in decision-making, related to the application of new knowledge. This is particularly urgent for developing countries. Governments and society at large should be aware of the need to use natural and social sciences and technology as tools to address the root causes and impacts of conflict.

Governments, through national science policies and in acting as catalysts to facilitate interaction and communication between stakeholders, should give recognition to the key role of scientific research in the acquisition of knowledge, in the training of scientists and in the education of the public government, private sectors, NGOs, people and should work in close collaboration and in a complementary manner to increase public understanding of science. [Declaration on science and the use of scientific knowledge... a new commitment, 1999 (Budapest)]

As mentioned in the Budapest world conference of science (1999), several major factors have transformed, and will continue to affect the relationship between science and society.

Although there is an urgent need to renew, expand and diversify basic science education for all, with emphasis on scientific and technological knowledge and skills needed to participate meaningfully in the society of the future on a broader scale, an increasingly scientifically oriented society needs science popularization in its widest sense, to promote an improved understanding of science and adequately orient public perceptions and attitudes about science and its application. So each country needs to have the capacity to design and implement its own science policy with responsibility within the global context.

A balanced development of a science base suitable for the country's needs requires an elaborate infrastructure and a stable institutional support, as well as the existence of an appropriate legal and regulatory framework. Regional and international networking and cooperation can facilitate the exchange of national experiences and the design of more coherent science policies (Introductory note to the science agenda – framework for action. Science for the twenty-first century: a new commitment, 1999).

History of Science-Iran

Iran has been a land of civilization, progress, industry and technology. Since ancient times, it has raised valuable people who have brilliantly shone in the world history. A study of the history of science in Iran will prove that Iranians has made great advances in scientific fields like botany, pharmaceutics, chemistry, zoology and mineralogy. Even now the Iranian scientists like ancient times, are making new advances in scientific fields. Iran has entered the scientific struggle and is availing of potentialities of its youth who have been pioneers in many scientific fields.

Continued qualitative and quantitative growth of international scientific productions of Iran is parallel to valuable achievements by Iranian. Researchers in science and technological fields have delineated a promising scientific outlook for the country (Mirrazavi, 2008).

On the other hand, following the latest bimonthly update to Essential Science Indicators (ESI) from Thomson Reuters, *sciencewatch.com*, has produced a listing of the scientists, institutions, countries and journals (March, 2010) that have achieved the highest percentage increase in total citations from the fourth bimonthly period of 2009 to the fifth bimonthly period of 2009 – that is from August 2009 to October 2009. These entities are singled out as rising stars in their respective field.

Iran has achieved rising star status in multiple fields, computer science, environment/ecology and molecular biology and genetics. Serbia and Iran have achieved rising star status in multiple fields. The full list of countries is given in the Table 9.1.

Table 9.1: The Countries Achieving the Highest per cent Increase in Total Citations for this Period.

Country	*Field*	*Country*	*Field*
UGANDA	Agricultural Sciences	**SERBIA**	Biology and Biochemistry
SERBIA	Chemistry	**SERBIA**	Clinical Medicine
IRAN	Computer Science	**LITHUANIA**	Economics and Business
SERBIA	Engineering	***IRAN***	Environment/Ecology
NIGER	Geosciences	**BULGARIA**	Immunology
SERBIA	Materials Science	**SERBIA**	Mathematics
IRAN	Molecular Biology and Genetics	**COSTA RICA**	Microbiology
CHAD	Multidisciplinary	**SERBIA**	Neuroscience and Behavior
CAMEROON	Pharmacology and Toxicology	**SERBIA**	Physics
PAKISTAN	Plant and Animal Science	**LEBANON**	Psychiatry/Psychology
ROMANIA	Social Sciences, general	**BOLIVIA**	Space Science

Science, Technology and Popularization of Science in Iran

Various agents are involved in popularization of science in Iran, each with different roles and functions, who need different types of training and incentives. The popularization of science requires integrated action by scientists, researchers, knowledge disseminators, NGOs, private sectors, journalists, Media, universities and members of scientific, social and research institutions. Their ultimate goal is to promote science and technology centers by creating training programs especially for science intermediaries, publish their scientific activities and proceedings in a simple style, creating national and international programs, include some related speeches in a special scientific occasion, and some other useful activities. They foster their procedures by coordinating, collaborating and making network; include all science centers, institutions and scientists to share their experiences, knowledge and skills.

Some of the important centers which cover the main activities of popularization of science in Iran are:

- ☆ National Research Institute for Science Policy (NRISP)
- ☆ Iranian Association for Popularization of Science
- ☆ Children's Book Council of Iran (CBC), etc.

The National Research Institute for Science Policy (NRISP)

The National Research Institute for Science Policy (NRISP) was founded in 1980 under the auspices of the Ministry of Science, Research and Technology (formerly the Ministry of Science and Higher Education). The appointments were based on the NRISP Constitution approved by the Ministry. Initially, NRISP consisted of six scientific groups, namely the Foresight group, the Popularization of Science group, the Science and Society group, the Science Policy group, the Economy of Science group and Scientometric group. These groups were headed by professors from the respective fields. The main mission of NRISP is acting as a think tank for the Ministry of Science, Research and Technology and preparing policy papers in Science and Technology Policy criteria for mentioned ministry in national level.

NRISP is a research institute that carries out basic, applied, and experimental research at national and international levels in order to develop processes and outputs of science and technology policy-making in the country and provides research and consultative services to policy-making organizations in the field of science and technology. The institute also publishes the research results to improve public understanding of science and research.

Vision

Given the approval and ratification of the "20- year vision of the Islamic Republic of Iran" and general policies of the fourth "five- year social, economic and cultural development plan" NRISP hopefully tries to play its due role as follows:

1. NRISP intends to have the highest rank (Center of Excellence) in the region (including; Middle Asia, Caucasus, Middle East and neighbors) in the field of science and technology policy- making in 2025 (20- year vision plan).
2. NRISP intends to be the best national think tank in the field of Science and Technology policy-making (STP) in 2013 (five- year vision plan).

Public Understanding of Science Research Department

This department was founded in 2006 to establish grounds for public understanding of science and its relevant policy- making in Iran.

Research Priorities

1. To design the national document of public understanding of science and public education including mission, policy guidelines and national standards;

2. To design and implement national network of public education of science and technology with the cooperation of public and private sector;
3. To perform research on the methods of promoting science and technology policy concepts among administrators and elites in the country;
4. To design and implement the program of continuous monitoring of public education and public understanding of science;
5. To carry out research on designing and implementing educational methods with the cooperation of people who promote science. (The National Research Institute for Science Policy (NRISP) website.

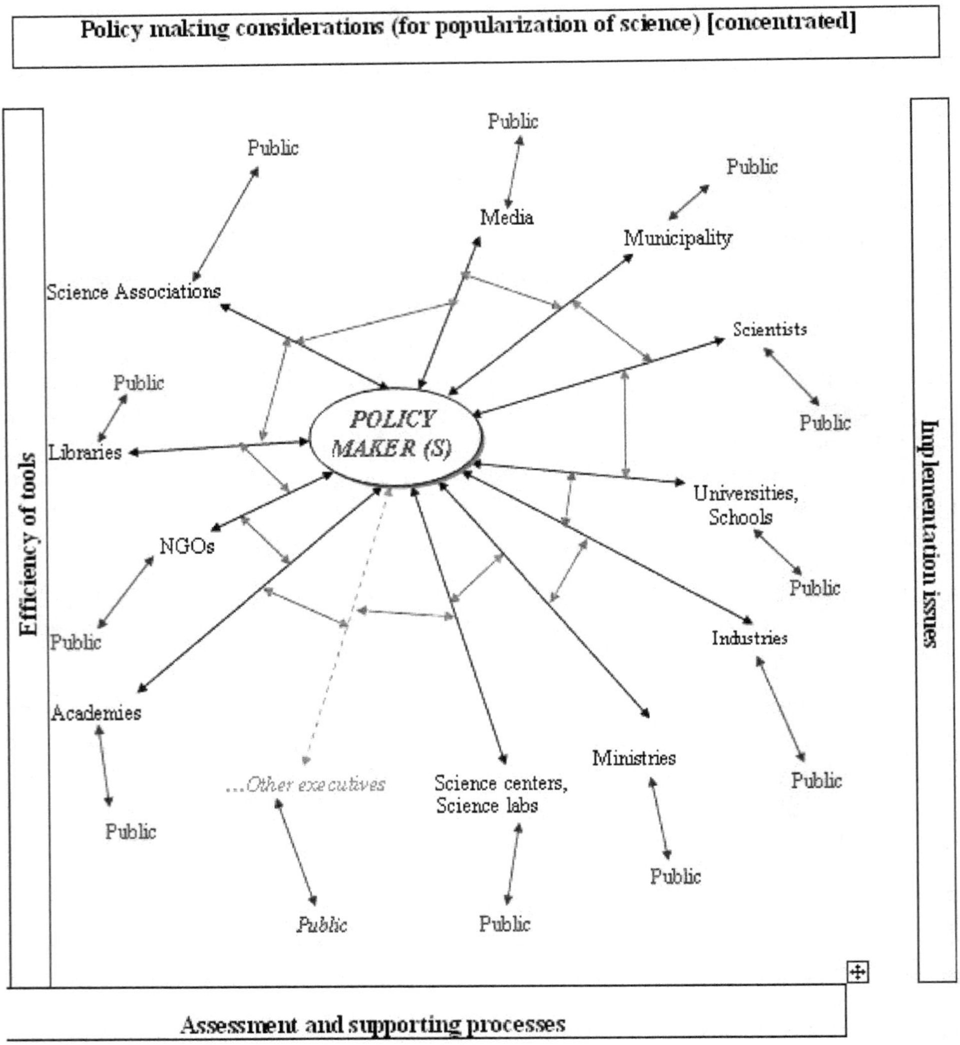

Figure 9.1: A National Prototype for Popularization of Science, Policy and Actions in Islamic Republic of Iran

A National Prototype for Popularization of Science, Policy and Actions in Islamic Republic of Iran

Due to the importance and necessity of preparing a national prototype for popularization of science in Iran, this national model is proposed.

A very simple mechanism of this national prototype is given below. As it is clear in this figure, all the relationships and fletches are two-sided and mutual. We do not have any up-down rules in this national prototype. Policy making, decision making, and actions are collaborative. In this mechanism, plans and actions are supported and get assessed through concentrated policy maker, or a team working of policy makers. All executives cooperate with a policy maker or policy makers (that consist of a group (of institutions or connoisseurs)) for planning, action (implementing) and other activities in a national scale. All executives in this national prototype collaborate with each other, too and the green fletches show this mutual relationship. Besides that, all executives have mutual relationship with public for popularization of their scientific achievements, for collaborating in special scientific programs, for planning their recent activities on popularization of science and etc.

In this mechanism, along with public understanding of science, public contributions in scientific processes get raised and this promotion in science understanding among public, foster the sustainable development processes in our country.

In this national prototype, popularization of science gets organized and all include organizations and public, collaborate on it in various types and scales. Besides, all executives make use of different kinds of tools and make benefit from efficiency of various tools for their plan in order to do their best. This national prototype is given below:

Conclusion and Suggestions

What is clear here is that people have right to know about the science and make use of it in their life to live better. Science is essential for sustainable development, so, Policy makers, decision makers and Governors' should know that along with science production, science popularization should be implemented in the country and needs more attention.

On the other hand, along with the long term planning, application of good technologies will determine future trends towards scientific approach as a national part of our daily life. So it is time to educate and teach the usage of these technologies to people, especially children.

References

The American Association for the Advancement of Sciences (AAAS) website. Available at: http://www.aaas.org/aboutaaas/

The American College Dictionary. 1963. Random House, Inc., New York.

Bauer, Martin W., Allum, Nick and Miller, Steve, 2007. What can we learn from 25 years of PUS survey research? Liberating and expanding the agenda. *Public*

Understanding of Science, 16: 79-95. Available at: http://pus.sagepub.com/cgi/content/abstract/16/1/79

Cornelis, Gustaaf c. Is popularization of Science possible? Available at: http://www.bu.edu/wcp/papers/scie/sciecorn.htm

Declaration on science and the use of scientific knowledge. Text adopted by the world conference on science: science for the twenty-first century: a new commitment. 26 June-1 July 1999. Budapest. Available at: http://www.unesco.org/science/wcs/eng/declaration_e.htm

Introductory note to the science agenda – framework for action. Science for the twenty-first century: a new commitment, 26 June-1 July 1999. Budapest. Available at: http://www.unesco.org/science/wcs/eng/intro-framework.htm

Kohlstedt, S.G., Sokal, M.M. and Lewenstein, B. 1999. The establishment of science in America. Washington DC: AAAS. Quoted in Martin W. Bauer, Nick Allum and Steve Miller: What can we learn from 25 years of PUS survey research? Liberating and expanding the agenda (Public Understanding of Science, 2007), 91. Available at: http://pus.sagepub.com/cgi/content/abstract/16/1/79

Making science and technology work for the poor and for sustainable development in Africa: parliamentary documentation, 2003. United Nations Economic and Social Council) (Third meeting of the committee on 2003. Sustainable development (CSD-3), 7-9 October). Available at: *www.uneca.org/*.

McGraw-Hill Encyclopedia of Science and Technology. 1960. McGraw-Hill Book Company, London, Vol. 12.

Mirrazavi, Firouzeh, 2008. *Scientific Advances of Iranians*. Available at: http://www.iranreview.org/content/view/663/37/

The National Research Institute for Science Policy (NRISP) website. Available at: www.nrisp.ac.ir

Redey, Soma. Communication technologies of science: a change of science-cognition. Available at: http://www.ifz.tugraz.at/index_en.php

Science watch website. Available at: http://sciencewatch.com/dr/rs/10mgr-rs/

The workshop on the popularization of science and technology, from office of science and technology organization of American states, Museu de Astronomia e ciências Afins, Ministry of science and technology, Brazil, 2004. Available at: http://www.science.oas.org/Components/Brasil/04_29_04_BRAZIL_Informe_Final_ENG.pdf

World Conference on Science: Science for the twenty-first century: a new commitment (An overview). 26 June-1 July 1999, Budapest. Available at: http://www.unesco.org/science/wcs/eng/an overview.htm

Chapter 10

ISC and Its Role in Science and Technology and Sustainable Development

Jafar Mehrad
President of Islamic World Science Citation Center,
Regional Information Center for Science and Technology,
Jam-e-Jam Ave, Jomhouri Eslami Blvd, Shiraz, Iran
E-mail: dean@ricest.ac.ir

ABSTRACT

This article deals with Islamic World Science Citation Center (ISC) as the first citation system in the Islamic countries and sustainable development. It attempts to describe the mapping of ISC the way it has been established. By establishing ISC and evaluating research performances of the Islamic countries, the weak points and strengths of researches carried out in Islamic countries could be identified. By presenting weak points and attempting for filling the gap by the Islamic countries, accessing sustainable development by them will be guaranteed. ISC considers all subject domains and so it is possible for this citation analysis system to identify capabilities and determining those domains which need more scientific activities. All operation done in ISC is based on research journals collected by Islamic countries.

Keywords: *Islamic World Science Citation Center, Research performance, Sustainable development.*

Introduction

In modern world, development of countries are dependent on forecasting the future needs of societies, reviewing their performance and goals, depicting their current status and presenting their future development path, in order to be able to satisfy the local, regional, national and international needs of the societies heading towards sustainable development.

To take these significant steps, determining the major pivots and priorities required for sustainable development are vital and inevitable. Each country has its own specific view and perspective towards operationalizing the development plans; however, the current universal negotiations vividly indicate the common attitude of all countries about "development based on science and technology". This approach is remarkably presented by the efforts of countries stepping towards knowledge – oriented economy. Universal experiences and successful development of industrialized countries also significantly prove that achievements of the developed countries is the result of going through science and technology path. In other words, today "science and technology" as the pivot of economical, social and cultural development, has turned into a common belief from the international point of view.

Iran, for organizing science and technology of Islamic countries has an outstanding and remarkable status. This position with regard to Islamic World Science Citation Center (ISC) is dedicated to Iran by Islamic countries. ISC by creating scientific capacities in a more extensive level, making innovations in science and technology domain, and assessing science in Islamic countries has contributed to sustainable development process in these countries.

Science and technology, is the most effective strategy of countries to confront the major challenges of the 21th century, the most outstanding of which is sustainable development.

In this article I will present the history and process of establishing ISC. Accordingly, it is believed that scientific productions, research journals, universities and research institutes that observe and assess the scientific development process of the countries and measure their growth rate are considered as the basic elements in sustainable development.

Before elaborating on the effect of ISC on sustainable development, I would like to present a brief review on citation analysis which shapes the basis of ISC programmes.

Citation analysis is the examination of the frequency, patterns and graphs of citations in articles and books. It uses citations in scholarly works to establish links to other works or other researches. It is one of the most widely used methods of Bibliometrics (Richard, 2004).

Along with peer review, it has over the past three decades been increasingly used to judge and quantify the importance of scientists and scientific research. Citation analysis is also the machinery behind journal "impact factors" –figures of merit that researchers take note of when deciding which journal to submit their work to so that it is read as widely as possible. Indeed, the output from citation studies is often the only way that non-specialists in governments and funding bodies – or even those in different scientific disciplines – can judge the importance of a piece of scientific research.

Researchers do a citation analysis for several reasons:

- ☆ To find out how much impact a particular article has had, by showing which other authors based some work upon it or cited it as an example within their own papers.

- To determine more about a field or topic by reading the papers those cite a seminal work in that area.
- To find out how much impact a particular author has had by looking at his/her total citations (Meho, 2007).

The basic idea or assumption behind citation analysis is that influential works or scientists are cited more often than others. In this sense, citations reflect the relative impact and utility of a work, author, department, or journal's publications within their larger scientific domains. It's a long time that citation analysis is applied as a tool for assessing the research performance of countries which are aimed at sustainable development.

Until recently, ISI (currently Thomsons Scientific) has been known as the only Institute evaluating the research performance of countries on the basis of citation studies. Then, SCOPUS a subsidiary of ELSEVIER, did also start in this regard and is now considered as one of the serious rivals of ISI. Moreover, Google Scholar is another scientometric tool which evaluates research output based on the scientometrics criteria. Parallel to the citation analysis activities of these institutes, some countries like China, have also taken steps to set up their national scientometric tools and have gained considerable achievements in this regard.

On the one hand, ISI and Scopus productions are internationally considered as powerful scientometric tools, but on the other, because of the ISI and Scopus bias towards English language, most of the scientific productions written in the national languages of non-English language countries do not have any opportunities to enter ISI or Scopus databases and in turn, will be kept out of the world scientists' sight. Therefore, Islamic World Science Citation Center (ISC) has begun implementing scientometric tools similar to those of citation systems in order to be able to evaluate the research performance of the Islamic world scientists and provide the grounds for implementing the Islamic world scientific network.

Historical Background and Importance of ISC

Extending from Mauritania to Indonesia, the Muslim world occupies the middle belt of the globe and assumes a position of extreme strategic importance.

A number of Islamic countries abound in natural wealth, others have greater potential of manpower and yet there are few which possess a relatively higher level of standards in education and research. These differences can foster the quality of close friendship and cooperation between these countries, especially when there is an identity of purpose, similarity of views, agreement in order of priorities, parallelism in the social and situational behaviour and above all, common faith and destiny. The hybridization of these elements can yield astonishing results in the creation of cooperative agreements and joint venture for quality training and research in the area of information collaboration.

One-fifth of the world populations are Muslims living in 57 countries. Among OIC member states, journals of 54 countries were analyzed based on the principle of availability. Figure 10.1 shows the distribution of journals in each country.

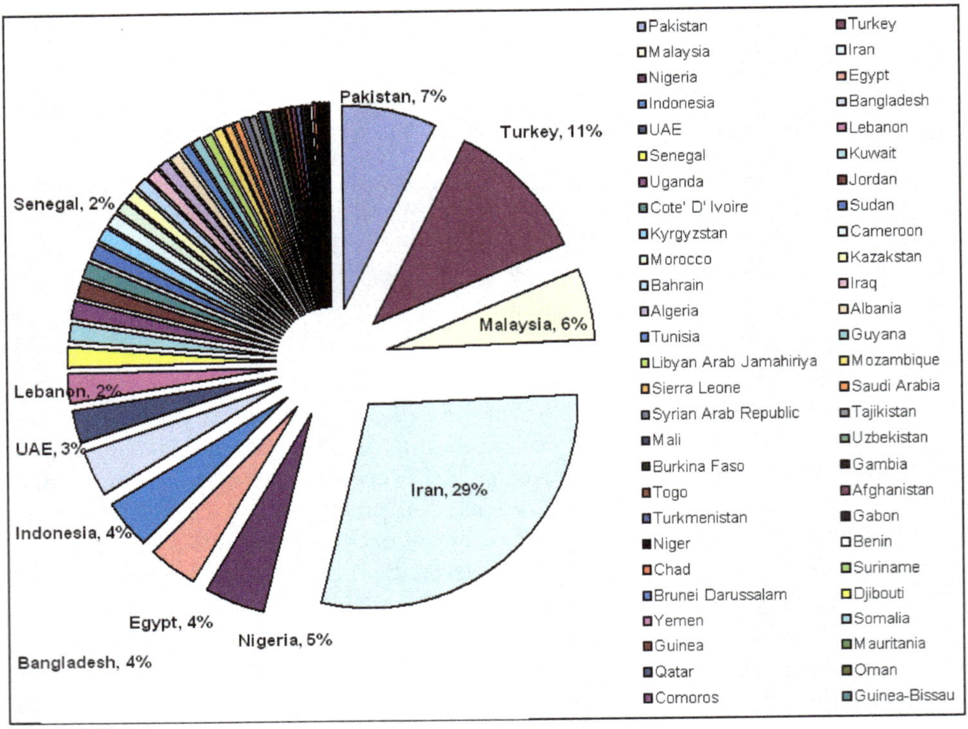

Figure 10.1: The Percentage of Research Journals in Islamic Countries

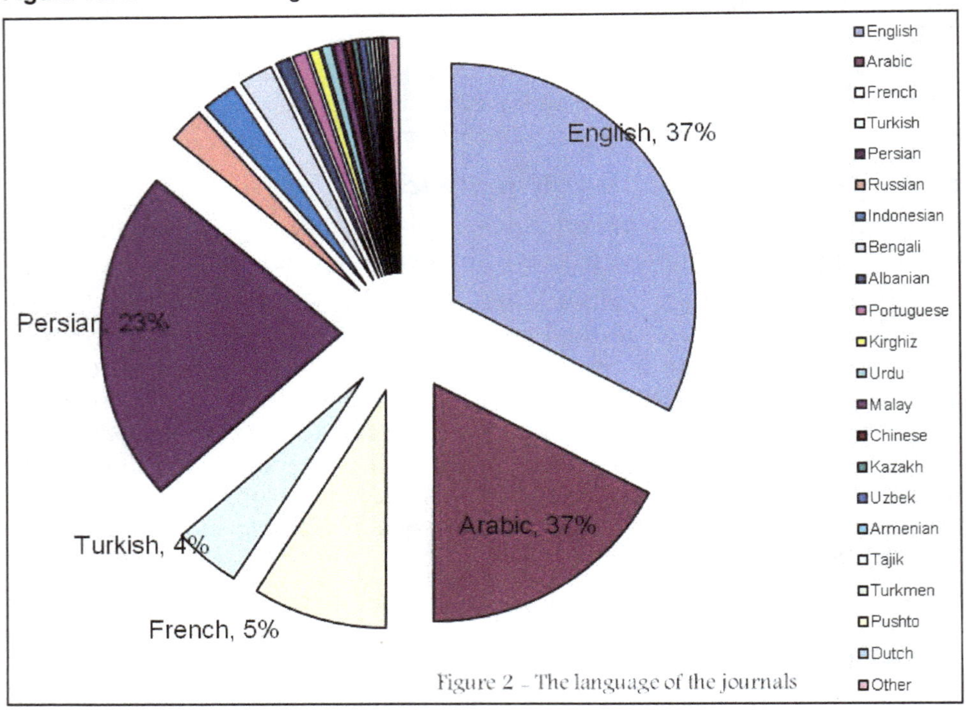

Figure 10.2: The Languages of the Journals Published by Islamic Countries

Figure 10.2 provides a language-based analysis of the scientific journals in the Islamic countries.

Studies done on the frequency of the publication of these research journals show that more than 50 per cent of the scientific journals are well qualified to be included in ISC.

Islamic countries should consider international indicators to be able to achieve sustainable development and enter to the world class universities and other ranking systems. To achieve this purpose, ISC is to process the Islamic countries research journals published in different languages. ISC mainly aims to evaluate institutes, countries, researchers and publications of the Islamic countries published in their national and native languages. ISC began its work on Persian journals at ISC in 2001 released its products in 2004. After ISC was accepted at 9th ISESCO General Assembly, Rabat, Morocco, ISC studied for the Arabic and English journals published by universities and higher education centers of Islamic countries in order to consider them for inserting in ISC.

Approval of ISC by ISESCO and appointing its policy plan by the Supreme Council for Cultural Revolution highlighted the mission of ISC on assessing research performance of the Islamic countries.

For being successful in ISC, the cooperation of the Islamic countries with ISC is of utmost importance.

There are relatively useful structures in the Islamic countries. ISC will help the structures to be developed. In this way, suitable grounds for the globalization of universities and research centers of the Islamic countries will be provided.

Now, there exist very few numbers of universities in the Islamic countries included in the world ranking from which we can mention Iran, Turkey, Egypt and Malaysia. Taking into consideration the research performance of the Islamic countries' universities based on the international criteria, will provide suitable grounds which make us expect more and more universities of the Islamic countries to be included in the world university ranking systems.

With regard to the importance of ISC, now ISI and SCOPUS are interested in signing MOU with ISC. For implementing such interest, there were made different meetings between ISC and Scopus in Netherlands and Iran which resulted in agreement on the MOU between ISC and Scopus.

Implementation Methodology

ISC selects research journals among those published by universities and research centers of the Islamic world. This selection criterion has been approved and is now being implemented on most of the Islamic countries.

Based on these regulations, hundreds of scientific journals from the Islamic counties in Farsi, Arabic, English and French have been acquired from 2000 on.

ISC began to work in 2001. Since then, about 1671 journals were chosen as the first batch of source journals in which the shares occupied by Basic Science,

Engineering and Technology, Medicine as well as Agriculture and Humanities, Arts and Social Sciences out of the numerous journals published in the Islamic countries.

The Impact of ISC upon Islamic Countries Scientific Community

ISC experts not only build and maintain this system but also have made some justifiable modifications including provision of full text papers of articles in Arabic, English and Farsi which differs from its counterparts such as ISI and Scopus.

Prompted by ISC evaluation of universities and research institutes, more and more organizations introduce incentives to encourage scientists to publish more papers in influential journals. Some higher education authorities in Iran and in the region remarked once and again that ISC has provided a solid foundation in the establishment of scientific journals, scientists, and researchers' views, in assessing universities and research centers and in comparing universities with each other. To be specific through ISC, more and more scientists in Iran and other Islamic countries have been increasingly aware of the importance to publish in domestic journals, so as to enhance effective scholarly exchange and establish themselves among the community. As a result, their papers productivity has been improved a great deal.

In short, ISC's capabilities can be summarized as following. Islamic countries can use these capabilities for their research progress and sustainable development:

Determining whether an article has been cited or not, and if yes, to what extent and by whom forms the major goal of ISC and its subsystems. It is supposed that the number of times an article has been cited by the experts of that field or other fields determines the significance of that article. In other words, the number of citations an article has received proves its national and universal value.

The more citations an article receives, the more significant the article is:

When the time span of receiving citation by an article is more, the article's dominant idea is also more durable.

Determining by whom and in which sources an article has been cited.

The more intellectual collaboration among scientists exists, the more reflection and dissemination of those scientists' thought in special media will take place.

Individuals in the same cores will publish and exchange their thoughts in the same media.

Possibility to study the impact of other fields and also on other subject areas (interdisciplinary relation).

Determining core scientists for perceiving their thoughts and establishing suitable relationship with them.

By carrying out ISC and controlling scientific production in ISC as well as applying scientometric criteria in assessing the scientific productions of the Islamic countries, one can find out about Islamic countries scientific atmosphere and scientists' performance in different aspects.

Approaching scientific growth stage and sustainable development will lead to changes in structures, regulations and investment. Accordingly, in ISC domain, providing positive grounds and carrying out scientific laws exactly to evaluate research performances of the countries will guarantee scientific progress and sustainable development.

I believe that by implementing ISC in Islamic world, ISC and its subsystems have turned into the tools for the growth of science.

As you noticed, ISC is the result of the significant efforts made by the government of Iran, Islamic countries and ISESCO. Our friends in nonaligned countries who are interested in joining ISC are most welcome.

References

Meho Lokman, I., 2007. The rise and rise of citation analysis. *Physics Worlds,* 20(1): 32–36.

Richard, E.R., 2004. *Foundations of Library and Information Science, 2nd edn.* Neal-Schuman, New York.

Chapter 11

Women Empowerment through Science and Technology Education

P.K. Gupta

Secretary General,

National Foundation of Indian Engineers (NAFEN), India

E-mail: nafenindia@nafenindia.com nafenindia@airtelmail.in

ABSTRACT

21st Century is unique in itself particularly with respect to fast changing technological developments and associated social, cultural and economic changes. Raising the economic levels of the society are major objectives of all the countries. These call for major initiatives for women to take up more jobs and thereby raising their purchasing power and income levels. This can only be possible if a large number of women enter science and technology area. It is interesting to note that at the start of the 21st Century, only 11 per cent of Fortune 500 companies have women on their boards and only two companies have women CEO's world wide.

Technically qualified women can also become self-employed (Entrepreneurship) thereby creating more job opportunities for fellow women. There is a strong correlation between female higher S&T education and several developmental indicators such as increased economic levels, productivity enhancement, improvement in health, marriage at more mature age, lower fertility etc.

In this paper the author discusses the present social and cultural, economic and technological environment prevailing world wide with respect to women taking up Science and Technology Education. Various impact factors both which motivate women to take up S&T education as well as the barriers for their entry are also highlighted.

Keyword: *Women empowerment; Science & Technology policy, Education.*

Introduction

Poverty alleviation and raising the economic levels of the society are major objectives of all the countries.

These call for major initiatives for women to take up more jobs and thereby raising their purchasing power and income levels of the household.

This can only be possible if a large number of women enter science and technology area may be education and subsequently employment.

It is interesting to note that at the start of the 21st Century, only 11 per cent of Fortune 500 companies have women on their boards and only two companies have women CEOs world wide.

Role of women in building up families and societies is undergoing vast changes particularly with the opening of economies and increased competition from MNC's.

The market is shifting from 'Sellers Market' to 'Buyers Market'. In this competitive environment, women have to prove their mantle in business organisations.

Technically qualified women can also become self employed (Entrepreneurship) thereby creating more job opportunities for fellow women.

Higher Science and Technology Education has a direct impact on women empowerment as it creates in them awareness about their rights, their capabilities and the choices and opportunities available to them.

There is a strong correlation between female higher S&T education and several developmental indicators such as increased economic productivity, improvement in health, marriage at more mature age, lower fertility, increased political participation and effective investments in the next generation.

Women empowerment will result in the acquisition of the essential knowledge and skills. It would enable one to engage in activities required for effective functioning of the individual in her group and community and use these skills towards the societal-economic development.

S&T Education, particularly of women, helps to control excessive population growth by promoting the concepts of family planning, collective health and well-being.

An educated family makes informed choices with respect to having a child as well as for maintaining its own and the children's health.

Thus educated families:

- ☆ Are more empowered and confident
- ☆ Have fewer children

- ☆ Have healthier and better educated children
- ☆ Are better equipped to enter the paid labour force, which is particularly important to the survival of female-headed households
- ☆ Enjoy higher levels of economic productivity
- ☆ Experience a longer and healthier life.

We can also say:

"Educate a man, you educate a person
Educate a woman, you educate a family"

Science Policy in India for Women

The main objective are:

- ☆ To provide an opportunity to women scientists for self-employment by utilizing their specialized domain knowledge in areas such as patenting, proof reading, science journalism, technical translation, clinical pathology labs, medical transcription etc.
- ☆ To create a large pool of trained women workforce with experience in the diverse areas mentioned above and building up a professional network.

The Constitution of India guarantees to all Indian women equality, no discrimination by the State, equality of opportunity, equal pay for equal work. In addition, it allows special provisions to be made by the State in favour of women and children, renounces practices derogatory to the dignity of women and also allows for provisions to be made by the State for securing just and humane conditions of work and for maternity relief.

In line with the above, Government of India has taken following steps to promote women empowerment through S&T Education:

- ☆ Flexible working hours for women with small children
- ☆ Funding for crèche facilities
- ☆ Construction of women's hostel blocks
- ☆ Annual research grants up to Rs 20 lakh for five years for young women scientists.
- ☆ Special fellowship scheme for women scientists to enable them to continue after a break in career as follows:
    - Scholarship for Research in Basic/Applied Science
    - Scholarship for Research in S&T–based Societal Programs
    - Internship for the Self-Employment
    - Special Women Scientist Cell set up in IITs, which holds gender sensitization workshops.

Prevailing Environment

- ☆ Knowledge and Information Based Society – Internet, E-Mail, E-Commerce, Multi Media, Virtual Reality, Information Super Highways, etc.
- ☆ Global Networking.
- ☆ High Mobility of Population from Rural to Urban Areas.
- ☆ Increasing Productivity.
- ☆ Reliance on Innovation and Creativity.
- ☆ Conscious Society – Customer aware of his Rights and Obligations.
- ☆ Role of WTO
- ☆ Emergence of Trade Blocks like EU/EC, SAARC, OPEC etc.
- ☆ Population Explosion
- ☆ Rapid Technological Changes
- ☆ Social Changes
- ☆ Literacy Levels
- ☆ Entire World as a Global Economic Village
- ☆ Transport Revolution
- ☆ Heavy Industrialization
- ☆ Fast Aging due to Hectic Life
- ☆ Fast Developing Political and Economic Changes In Eastern Europe
- ☆ Shift of Manufacturing Bases to Nations like India, South East Asia, etc.
- ☆ Enhanced Medical Facilities
- ☆ Multi Media, Information Super Highways.

S&T Education

Table 11.1: Streams and Disciplines of the S&T Education

Sl.No.	*Streams*	*Disciplines*	*Level*
1	Science	Physics, Chemistry, Maths, Zoology, Botany	B.Sc., M.Sc., Ph.D
2	Engineering and Technology	Electrical, Electronics, IT, Computers, Automation, Bio-technology, Nano-technology, Telecommunication	B.E./B.Tech., M.E./M.Tech., Ph.D

Table 11.2: Typical Distribution in a Engineering and Technology Institute in a Metro City (Delhi) in India

Disciplines	*Periods/Genders*									
	2007-08		*2008-09*		*2009-10*		*2010-11*		*Overall per cent*	
	Male %	*Female %*	*Male %*	*Female %*	*Male %*	*Female %*	*Male %*	*Female %*	*Male %*	*Female %*
Information Technology	80.30	19.70	70.18	29.82	84.13	15.87	65.15	34.85	74.94	25.06
Computer Science Engineering	90.08	9.92	69.37	30.63	87.88	12.12	87.02	12.98	83.59	16.41
Electronics and Computers	93.13	6.87	92.66	7.34	88.46	11.54	93.13	6.87	91.85	8.15
Manufacturing Automation Engineering	98.45	1.55	99.13	0.87	100.00	0.00	100.00	0.00	99.40	0.60
Electrical Engineering	84.13	15.87	82.69	17.31	67.27	32.73	84.13	15.87	79.55	20.45

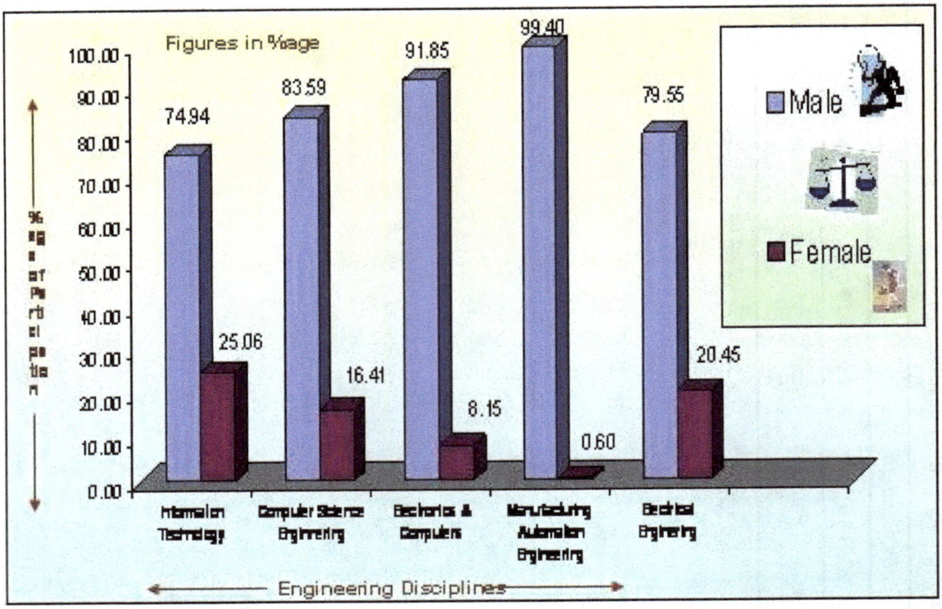

Figure 11.1: Overall Distribution of 4 Years Period (2007-2011)

How to Motivate Women to come forward for S&T Education

Let us see what are the various Impact Factors. They must be identified very carefully and examined critically (CSF and KRAs) and at the same time barriers for their entry should also be identified and suitably taken care of.

Table 11.3: Impact Factors

Factors	*Elements*
Social and Cultural	☆ Age at the time of admission at Bachelor, Master, Doctorate Level
	☆ Background – Rural/Urban—Regional Imbalances
	☆ Competition for admission at the time of entry in the courses
	☆ Influence of parents and elders
	☆ Religion
	☆ Size of family
	☆ Education background of the parents
Economic	☆ Employment Opportunities
	☆ Impact of Entry of MNC's on employment opportunities in the country
	☆ Income of Parents
Technological	☆ Aptitude for higher S&T education
	☆ Use of Internet and E-mail
	☆ Any other Technological Development

Contd...

Table 11.3–*Contd...*

Factors	*Elements*
General	☆ Few single sex institutions, ☆ Inadequate supply of female faculty ☆ Insufficient hostel facilities ☆ No reservation of seats ☆ Lack of special admission procedures/criteria ☆ Lack of special scholarships and financial assistance ☆ Improper counselling and guidance ☆ Cultural restrictions and gender bias in employment ☆ Inadequate links between the programs offered and the opportunities for the women in the employment ☆ Gender bias in curricula ☆ Early marriages ☆ Would flexible education schedules help females fit in their other tasks? ☆ Do girls/women have access to safe transport? ☆ Will creating facilitating centres in each institute with female councillors help?

Note: Scores can be allotted for each factor in order to arrive at the overall impact. This will also give relative positioning of each impact factor.

Conclusions

Many countries in comparison with the developed countries and also some of the developing countries suffer from a malady of low levels of S&T education among the women population. As a result, the development potential of these countries is substantially less and not fully utilised. This further leads to poor quality of life, slow economic and social growth, poverty enhancement etc. It is therefore high time that the Policy Planners and Decision makers evolve policy both short-term and long-term which encourage women to take S&T education as their career. It will lead to following benefits:

☆ Poverty alleviation by increasing house hold income and purchasing power
☆ Removal of gender Inequalities
☆ Better quality of life
☆ Making women self reliant
☆ Acceleration of economic growth and social progress
☆ Enhance enterprise development and entrepreneurship
☆ Participation of more women in leadership and decision making

Acknowledgement

The author is greatly obliged to Prof. Dr Arun P Kulshrestha, Director and CEO, NAM S&T Centre, New Delhi for his invaluable advice and suggestions for preparation of this paper. The author would also like to thank Mr M. Bandyopadhyay, Senior Expert and Administrative Officer, NAM S&T Centre for his advice.

Chapter 12

The Role of Science and Technology in Biodiversity based Sustainable Development

Palpu Pushpangadan*, Ijinu, Vipin Mohan Dan and V. George
Amity Institute for Herbal and Biotech Products Development, India
**E-mail: palpuprakulam@yahoo.co.in*

ABSTRACT

Scientific Knowledge and productive technology are highly objective and therefore leads to continuous innovation and improvement. Scientific knowledge is always variable because it is based on careful observations of matter/the natural process based on which conformations and conclusions are made. It excludes all unsubstantiated hypothesis and observations. Experimentations are the methods of science. Any new knowledge adds to the existing body of knowledge. There is nothing like a finite knowledge in science. It is an ever-growing dynamic system and perhaps it is the characteristic attribute of science. The environment of living organisms in general and the humankind in particular is always in a flux of change. The organisms need to continuously adjust and adapt to lead a healthy and satisfying life. The human kind is depending on Biodiversity for its sustenance. Modern scientific methods and technology can contribute a great deal in conservation and sustainable utilization of Biodiversity. Biotechnology could be profitably employed to undertake the study of structure and functional dynamics of every organism and also for conserving and optimal utilization of resources in sustainable manner. *In vitro* conservation in the form of seed banks, tissue repositories, pollen/ovary bank or even DNA bank are some of the modern biotechnological methods of conservation of rare and endangered species besides other methods like *in situ* conservation or other garden based *ex situ* conservation, etc. Biotechnological tools can also be employed in manipulating the living organism to make and modify new value added products thereby converting biodiversity into economic wealth. Such advancement triggered a paradigm shift of world trade and economic activities from 'resource based economies' to 'knowledge-based economies'. A new thinking centered on the concept of 'knowledge engineering' for building up future 'knowledge societies' and 'knowledge industries' is now gaining attention and acceptance all over the world. Generating new knowledge and converting it into useful value added products, production process and services using modern science and technology, and subsequent transfer of such products and technologies to industry and commerce with appropriate safe guard measures for Intellectual Property Rights (IPR's) protection are some of the key strategies required for third world countries.

***Keywords**: Scientific knowledge, Technology, Biodiversity, Biotechnology, Conservation, IPR.*

Introduction

From the very beginning of the human civilization human communities living in different agroclimatic conditions have acquired unique knowledge about their ambient biodiversity by inherent instinct, or intuition, or by accidental discovery, or by error, or empirical reasoning or even by conscious trial and experimentation. The above said processes enabled him to select a variety of natural resources, more particularly the biological resources to meet his primary needs like food, nutrition, medicine, shelter, clothing and to build his ever growing material culture and civilization. Many human communities later established advanced material culture and began to live in towns and cities. But majority of the people in the third world countries continue to live in and around forests or in villages. While the advanced societies developed written languages and documented their knowledge in written records and the traditional communities maintained their knowledge system through oral traditions.

The mechanisation and undue objectification of human life and health care systems of the present era have culminated to an extreme nexus between physician and patient by the interpolation of a third entirely mechanical thing, the machine, replacing the creative synthesizing role of the traditional physicians. This has resulted in the dehumanization of the medical system. In contrast to this scenario of the modern medicine, the traditional medicine attempts to embody a holistic approach *i.e.* viewing an individual in his totality within society and the ecological environment. It emphasizes the view point that ill health or disease is brought about by an imbalance or disequilibrium of man's physiological, psychological, behaviourial, ecological and spiritual environment and not just by an external pathogenic agent, be it a micro-organism or otherwise. No doubt, the modern medicine has accomplished great strides in developing many new life saving drugs. Modern health care system stressed more on the curative and to a lesser extent to the preventive aspects of diseases and very little has been done on the health promotive aspects. Problems of health have been replaced by problems of drugs and diseases. Instead of medicine for man, we have men for medicine. Modern medicine may help man to provide apparent physical health but is devoid of mental, social and spiritual health. Modern medicine is more concerned with the cure of disease but remains indifferent to health preservation. It is in this context that the relevance of the holistic approach of the traditional health care practice becomes important. There have been an ever increasing production and consumption of phyto-medicines based on various traditional systems of medicine both in the developing and developed countries. A steady global market for many such herbal products is emerging.

New technologies are constantly being developed to isolate and identify the components responsible for the activity of these plants. But these technologies should consider and possibly use the fact that the biological activity of plant extracts often results from additive or synergistic effects of its components. Another possibility is the qualitative and quantitative variations in the content of bioactive phytochemicals, which are currently considered major detriments in their use as medicine. Different stresses, locations, climates, microenvironments and physical and chemical stimuli, often called elicitors, qualitatively and quantitatively alter the content of bioactive secondary metabolites. Enzymatic pathways leading to the synthesis of these phytochemicals are highly inducible (Ebel and Costa, 1994). This is particularly true for phytochemicals that are well documented for their pharmacological activity, such as alkaloids (Facchini, 2001), phenylpropanoids (Dixon and Paiva, 1995) and terpenoids (Trapp and Croteau, 2001) whose levels often increase by two to three orders of magnitude following stress or elicitation (Darvill and Albersheim, 1984). Thus, elicitation-induced, reproducible increases in bioactive molecules, which might otherwise be undetected in screens, should significantly improve reliability and efficiency of plant extracts in drug discovery while at the same time preserving wild species and their habitats. Standardization, optimization and full control of growing conditions should guarantee a cost-effective and quality-controlled production of many plant-derived compounds. This kind of standardization and quality control of the plant-based drugs will improve safety of these drugs and promote their usage.

Convention on Biological Diversity and its Implications

The UN Convention on Biological Diversity (CBD) explicitly recognises the sovereign rights of the States over their biological resources (Article 15{1}). CBD also enjoins the States to respect, preserve and maintain the knowledge, innovations and practices of indigenous and local communities, and share with them the benefits arising from the utilization of such knowledge, innovations and practices (Article 8j). Since the Traditional Knowledge Systems (TKS) are threatened both by the national and international developments, and the benefits derived from the usage of TKS transcend national boundaries, it is imperative that any set of actions to implement CBD will have both national and international implications. CBD also recognizes such inherent impediments. Under Article 15(1), it stipulates that the authority to determine access to genetic resources rests with the national governments and is subject to national legislation. Article 15(4) and Article 15(5) subject this access to other contracting parties on 'mutually agreed terms' and 'prior informed consent'. If we read the Articles 8 and 15 together, we will get an implicit impression that the national governments have to play the key role in the implementation of these articles by due process of legislation and formulation of appropriate rules and regulations.

CBD has been founded upon three main objectives: (*i*) conservation of biological diversity, (*ii*) the sustainable use of its components, and (*iii*) fair and equitable sharing of the benefits arising from the use of genetic resources, including by appropriate access to genetic resources and by appropriate transfer of relevant technologies. While CBD dwells upon the sovereign rights of nations over their biological diversity, the TRIPs provide for an effective and adequate protection of IPRs of individual or corporate bodies for new inventions technologies and products developed by them.

The fast advancements made in science and technology during the 20th century led to the technological breakthroughs in Information Technology, Biotechnology- more particularly in molecular biology and genetic engineering/metabolic engineering (genomics, proteomics, metabolomics)–and Herbal Technology. Such advancements triggered a paradigm shift of world trade and economic activities from 'resource-based economies' to 'knowledge-based economies'. A new thinking centered on the concept of 'knowledge engineering' for building up future 'knowledge societies' and 'knowledge industries' is now gaining attention and acceptance both nationally and internationally. Generating new knowledge and converting it into useful value added products, production process and services using modern science and technology, and subsequent transfer of such products and technologies to industry and commerce with appropriate safeguard measures for Intellectual Property Rights (IPR's), protection are of some of the key strategies required for third world countries.

CBD explicitly recognises the sovereign rights of the States over their biological resources (Article 15{1}). CBD also enjoins the States to respect, preserve and maintain the knowledge, innovations and practices of indigenous and local communities, and share with them the benefits arising from the utilization of such knowledge, innovations and practices (Article 8j). Since the traditional knowledge systems are threatened both by the national and international developments, and the benefits derived from the usage of TKS transcend national boundaries, it is imperative that any set of actions to implement CBD will have both national and international implications. CBD also recognizes such inherent impediments. Under Article 15(1), it stipulates that the authority to determine access to genetic resources rests with the national governments and is subject to national legislation. Article 15(4) and Article 15(5) subject this access to other contracting parties on 'mutually agreed terms' and 'prior informed consent'. If we read the Articles 8 and 15 together, we will get an implicit impression that the national governments have to play the key role in the implementation of these articles by due process of legislation and formulation of appropriate rules and regulations.

CBD expects the member countries to evolve strategies for conservation and sustainable utilization of biodiversity also an equitable sharing of benefits accrued from the use of biodiversity and associated knowledge system. However, another international regime, the WTO which came into force in June 1994 with another agenda which is almost in collision with the CBD. The Trade Related Intellectual Property Rights (TRIPs) and other regulations contained in the WTO are not conducive to conservation and sustainable utilization of biodiversity more particularly the equitable sharing of to the welfare of humankind. The countries of the world are now debating and trying to develop a workable strategy that bridge CBD and WTO so that the precious biodiversity could be conserved and its fruits could be utilized in a sustainable manner in the best interest of the health of the environment and prosperity of the whole mankind.

Implication in Traditional Medicine

Thus there are more challenges ahead for the Traditional Medicine practitioners and the researchers working in this area should make this system of medicine a more acceptable one and more standardized one. This can only be achieved if all the stakeholders of Traditional Medicine make a concerted effort with the proper backing of the governments to globalize and popularize this system of medicine. It is only by fusing ancient wisdom and modern science we can achieve it. Dr. Mashelkar while making recommendation after a national seminar on Ayurveda held at Chitrakoot in 2005 suggested a Golden Triangle Traditional Medicine/knowledge of modern science and modern medicine to raise the status of Ayurveda s a global health care system that could provide life long healthcare for humankind. An earnest step in this regard has already been taken as a joint effort by Council of Scientific and Industrial Research (CSIR) under the dynamic leadership of the then Director General, Dr. R. A. Mashelkar. CSIR was operating a twenty laboratory networked programme on discovery of bioactive molecules from medicinal plants used in the Indian traditional systems of medicine. It was for the first time in the history of CSIR that it had forged relationship with Indian traditional systems of medicine, particularly Ayurveda and Siddha. It was started in mid-1990s when CSIR signed the MOU with Arya Vaidya Sala in Kottakal. Dr. Valiathan was indeed instrumental in bringing CSIR and Arya Vaidya Sala together. While signing the MoU at Kottakkal, Dr. Valiathan said "This is a holy place where two rivers are meeting; the river of traditional knowledge and ancient wisdom represented by Arya Vaidya Sala and the river of modern knowledge represented by CSIR". Currently under CSIR's New Millennium India Technology Leadership Initiative (NMITLI) programme, several scientists have been working on establishing pharmacoepidemiological evidence-base to Ayurvedic medicine, practices and development of standardized herbal formulations. Randomized controlled clinical trials for rheumatoid and osteoarthritis, hepatoprotective, diabetes, hypolipidemic agents and Parkinson's disease and many others have reasonably established clinical efficacy. A review of some of the exemplary evidence-based and approaches have been also made by Mashelkar. He also stated "It is clear that we must increase our presence internationally through well researched publications related to quality, safety and efficacy of Ayurvedic medicines in international peer-reviewed journals. The Ayurvedic, pharmaceutical and medical scientists must come together to make this possible".

A significant aspect of the traditional medicine, which is mostly location specific – 'Local Health Traditional' (LHT) is that it is self reliant in nature and is almost entirely autonomous in character and rooted deep in the community, social traditions and cultural values. Indeed this was the reason for WHO to recognize the intrinsic importance of plant based traditional medicine while declaring its goal of health for all by 2000. WHO emphasize the strategic role of traditional medicine in providing primary healthcare needs of the rural population of the world. Herbal medicine has found its resurgence by the close of the 20th Century. There has been now an ever-increasing demand for plant-based medicine in world over. This is mainly due to the increasing realization of the harmful side effects as well as the high cost of modern medicine. Over 80 per cent of the medicinal plants are collected from wild sources.

But the increasing demand has led to over exploitation as well as unscientific extraction. Also the habitat degradation due to human interaction in the forest is also causing the rarity and even extinction of many important medicinal plants. Therefore conservation, domestication and commercial cultivation of medicinal plants have become an urgent need of the day. Cultivation of medicinal plants is also essential to ensure quality of raw drugs, a pre-requisite for quality, control and consistency in the manufacture of the finished herbal drugs. Cultivation and its processing of medicinal plants also can provide gainful mass rural employment. It is reported that in China about 100 million people are involved in medicinal plant industry. In India at present it is hardly 10 million. Considering the rich and validated biodiversity and the great traditional medicinal heritage, it is possible for India to involve rural people in massive scale in cultivation and processing of medicinal plant and to enable the plant based pharmaceutical/drug industries to produce quality herbal drugs and successfully play a major role in the global market of herbal drugs and plant based pharmaceuticals and cosmaceuticals. CSIR laboratories like National Botanical Research Institute (NBRI), Central Institute of Medicinal and Aromatic Plants (CIMAP), Central Drug Research Institute, Indian Institute of Integrative Medicine (Regional Research Laboratory), Jammu have highly focused programmes in herbal drugs and plant based pharmaceuticals and allied product development. Central Council for Research in Ayurveda and Siddha (CCRAS) and Indian Council of Agricultural Research, some universities and various other R&D organizations in the private sector are also undertaking diverse programmes in developing protocols for cultivation post-harvest handling, standardization and quality control of medicinal plants and its products. Government of India has also established a National Medicinal Plant Board (NMPB) for promoting the medicinal plant cultivation and its sustainable utilization. Still more concerted and coordinated efforts by all these actors are required to enable India to become a front runner in this fast developing sector of Industry.

Scientific Basis for Sustainability

Modern scientific methods and technology can contribute a great deal in conservation and sustainable utilization of biodiversity. Biotechnology could be profitably employed to undertake to study the structure, function and dynamics of every organism and also in conserving and optimal utilization of resources in a sustainable manner. *In vitro* conservation in the form of seed banks, tissue repositories, pollen/ovary bank or even DNA bank are some of the modern biotechnological methods of conservation of rare and endangered species besides other methods like the *in situ* conservation or other garden based *ex situ* conservation, etc. Biotechnological tools can also be employed in manipulating the living organisms to make and modify new value added products thereby converting biodiversity into economic wealth. Biotechnology is already playing a significant role in converting biodiversity into industrial and commercial valuable products, processes having increased productivity and application in many crucial areas such as agriculture (including Aquaculture), health care (medicines, vaccines, diagnostics, gene therapy), environmental protection, Bio-energy, etc. The emerging area of biotechnological application on biodiversity is known as bioprospecting. Nicolas Mateo *et al.* (2001) describes bio-prospecting as the "systematic search for genes, natural compounds, designs and whole organisms in wild life with a potential for product development by biological observations and biophysical, biochemical and genetic methods without disruption to nature". Bio-prospecting has thus three facets like "chemical prospecting, gene prospecting and bionic prospecting".

Purity, batch to batch consistency in quality of medicinal plants are some of the critical requirements of botanicals. The most important things for assuring quality of botanical medicines require preparation of a comprehension passport description of the species. This include correct identification of the plant species, selection of the correct genotype of the species, right edaphic and climatic conditions for its growing and right stage at which the harvesting of the medicinal parts to be made and proper post-harvest handling etc. A well defined protocol and standard operation procedure (SOP) from selection of the right genotype, cultivation methods, harvesting, post-harvest handling, preprocessing, storage and upto manufacturing need to be worked out and meticulously followed for herbal drugs. It may be noted from the figure the various requirement of sourcing medicinal plants. This include correct taxonomic identification and authentication, study on the medicinal part: root, stem, bark, leaves, flowers, fruits, nuts, gum, resins etc., collection details: Location, stage and developmental stage or growth of the plants for collection methods, pre-processing if any, storage etc. (Pushpangadan and Govindarajan, 2003 and 2005). This is followed by the organoleptic test of raw drug *i.e.* evaluation by means of sensory organs: touch, odour taste, microscopic and molecular examination, chemical composition (TLC, GLC, HPLC, HPTLC, GC, AAS, capillary electrophorosis, DNA fingerprinting), biological activity of the whole plant, and shelf life of raw drugs. This is followed by well defined Good Manufacturing Practices (GMP) and scientific validation including toxicity evaluation, chemical profiling, pharmacodynamics – effect of drug in the body, pharmacokinetics – absorption, distribution, metabolism, mechanism of action and execution, proper dosage form, proper presentation and packing and proper claim of therapeutic merits – Compared with other drugs.

Global Strategy for Plant Conservation

The alarming concern over the increasing loss of plant diversity across the world has resulted in the development of several national, international and global action plans, strategies, policies and legal frameworks – all aiming at halting the current and continuing unacceptable loss of plant biodiversity by way of supporting and facilitating appropriate plant conservation initiatives at all possible levels. *The Global Strategy for Plant Conservation (GSPC)* is one of such recent initiatives, adopted by the Conference of Parties (CoP) to CBD at its sixth meeting held in April 2002 in the Hague. GSPC includes a series of 16 out come – oriented targets to be achieved by the year 2010. All the 16 targets of GSPC ultimately aim at practical conservation of threatened plant diversity through *in – situ, ex – situ* or a combination of both methods, besides enhancing the capability of plant – rich countries in undertaking integrated conservation methods and strategies, including recovery and reintroduction research along with building up improved awareness and education on conservation and sustainable use of plant resources among the public, students, teachers, administrators,

NGOs and other stakeholders. Targets 3, 8 and 16 of GSPC are of direct relevance to botanic gardens and other *ex – situ* conservatory centres as they aim at:

1. Development of models with protocols for plant conservation and sustainable use, based on research and practical experience (Target 3, GSPC);
2. 60 percent of threatened plant species in accessible *ex – situ* collections, preferably in the country of origin, and 10 percent of them included in recovery and restoration programmes (Target 8, GSPC);
3. Networks for plant conservation activities established or strengthened at national, regional and international levels (Target 16, GSPC).

CoP to CBD further requested the Subsidiary Body on Scientific, Technical and Technological Advice (SBSTTA) to develop ways and measures within CBD's thematic and cross – cutting programmes of work for promoting implementation of GSPC and for monitoring and assessing its progress. The CoP VI Session particularly mentions about the role of botanic garden in conservation and sustainable utilization as well as in eco – education/environmental education in botanic gardens. Several countries, including India have initiated relevant programmes and activities for implementation of the GSPC targets at national level.

The National Biodiversity Strategy and Action Plan (NBSAP 2002, MoEF) of Govt. of India identifies *ex – situ* conservation of plant diversity as a priority area of action and it suggests strengthening and enhancing the role of botanic gardens, home gardens and other *ex – situ* conservatory networks in India as important ways and means to achieve the above goals. The Ministry of Environment and Forests (MoEF) is implementing a special assistance programme to support selected botanic gardens in India to equip them with adequate infrastructure and other resources to support '*ex – situ*' conservation of plant diversity.

In 2002, the Botanic Garden Conservation International (BGCI), U.K. with financial support from one of the world's largest financial services organization – the HSBC, has launched an eco – partnership programme "Investing in Nature" – with the global mission of conserving threatened plants for ensuring the prosperity of future generation and local community through an effective world – wide network of botanic gardens. BGCI operates this botanic garden – based global network programme in 11 countries such as Argentina, Brazil, China (including Hong Kong), India, Indonesia, Japan, The Middle east, North America (including Canada and the USA) and South – East Asia. With Dr. P. Pushpangadan, then Director, National Botanical Research Institute (NBRI), Lucknow was the implementing agency for the 'Investing in Nature' Programme in India (IIN – India) and through this initiative NBRI organized a number of activities to support the partner botanic gardens in India to implement various conservation, education, databasing, networking, training, capacity building, and community empowerment programmes. The main focus of this five-year programme (2002 to 2006) was the implementation of the International Agenda for Botanic Gardens in Conservation (Wyse Jackson and Sutherland 2000) and the Global Strategy for Plant Conservation (GSPC).

One of the significant activities under '*Investing in Nature – India*' was to implement a *Small Grant Project Scheme* through awards of several grants per annum for 1 to 2 years in support of the best Indian botanic garden plant conservation projects that could contribute to biodiversity conservation, environmental education, and sustainable development initiatives for the benefit of the local communities. Under this new initiative, the Indian botanic gardens were poised to accept the challenge of conserving as many as 600 rare and threatened species of India's flora and making them in accessible *ex – situ* collections for recovery and restoration programmes. Since 2003, 24 institutions representing botanic gardens, arboreta and community/conservation organizations under Central and State Government Departments, Universities and NGOs had been supported with small grant projects to undertake their activities, relevant to *ex – situ* conservation and reintroduction research.

NBRI is now also a national and international referral centre for biodiversity conservation, sustainable utilization, bioprospecting, biotechnology and molecular biology with accreditions to several National and International agencies/ organizations such as CBD, UNEP, WHO, IUCN, BGCI, MoEF, DBT, DST, etc. With its existing R and D expertise in plant conservation research and allied areas and with its recognition as a national coordinator of such an important global eco – partnership programmes as the *Investing in Nature – India (BGCI – HSBC – NBRI collaboration)*, NBRI is well placed to function as a model of a Lead Botanic Garden. Such a garden should have facilities for conservation and germplasm augmentation cum training centre for the Rare, Endangered and Threatened (RET) plant – species and economically and ecologically important (Flagship species, Keystone plant species, Indicator taxa etc.) plant resources of the Gangetic Plains. NBRI can certainly help in establishing such lead gardens. It is important to establish at least four to start with and later 10 such lead botanic gardens representing the ten biogeographic zones of India is planned.

The unknown potentials of genes, species and ecosystems represent a never ending biological frontier of inestimable value. Many materials for human needs, other than foods such as medicines cosmetics and perfumery are obtained from complex organic chemicals mostly referred as secondary plant metabolites of plants. According to a rough estimate 20 to 30 per cent *i.e.* 50,000 to 80,000 species of higher plants particularly those found in tropical and subtropical regions of the world have aromatic and medicinal value. The medicinal and aromatic properties of these plants are due to the presence of various simple and complex organic compounds generally known as secondary metabolites of all the biomes, tropical forests contain the largest portion of plants yielding secondary metabolites. Tropical plants presumably synthesizes these unique biocompounds for defense, offence, protection and adjustments with the local environment.

The factors influencing phyto-chemical profile of medicinal plants are:

- ☆ The therapeutic effect of medicinal plants are due to the presence of the biologically active chemicals which are mostly secondary metabolites such as alkaloids, glycosides, saponins, coumarins, flavonoids, terpenes, etc.

- 75 per cent to 80 per cent of the medicinal plants required in Traditional Medicine (TM) is collected from wild.
- 35 per cent to 40 per cent of the medicinal plants used in TM are cross-pollinated species and hence there exists vast – intraspecific genetic variability with respect to the morphological and chemical constituents.
- Genetic variants (at Ploidy level or gene level) leading to the variability in the chemical composition.
- Geographical and edaphic factors-altitude, soil composition, microbial load/association, climate, temperature etc.
- Seasonal changes (rainfall, photoperiodism, drought, water stress, etc.)
- Seasonal variations for eg. the alkaloid composition in the leaves of *Adhatoda vasica* has been recorded. It is reported to be as lowest in the months of February and March and highest in the months of August, September and October. Similarly variation of alkaloid contents based on the age of the plant is recorded in *Hoarrhena antidysenterica.*
- Association patterns including animals and insects.
- Lunar period.

With the revival of natural products there has been a tremendous increasing in preference for herbal drugs, cosmetics, natural flavors, perfumes and dyes and pharmaceuticals as they are found to be comparably harmless and have less effect. Thus there is unprecedented demand of the forest resources leading the over exploitation and even biopiracy.

DNA based molecular markers are proving to be a versatile tool in the plants genome analysis and in differentiating different genotypes. Various techniques like Random Amplified Polymorphic DNA (RAPD), Fragment Length Polymorphic DNA (AFLP), Restricted Length Polymorphic DNA (RFLP) and Inter Simple Sequence repeat (ISSR) etc. are now being successfully used for such genetic analysis of medicinal plants and also for the characterization of semiprocessed and even fully processed herbal products. DNA markers are highly stable and specific. It has immense applications in the standardization of medicinal plants and its products.

Thus Science and Technology have a key role in Biodiversity based sustainable development.

Acknowledgement

Authors are thankful to Dr. Ashok K. Chauhan, Founder President, RBEF for providing the facilities and encouragement.

References

Darvil, A.G. and Albersheim, P., 1984. Phytoalexins and their elicitors: A defense against microbial infection in plants. *Ann. Rev. Plant Physiol. Plant Mol. Biol.*, 35: 243–275.

Dixon, R.A. and Paiva, N.L., 1995. Stress induced phenyl propanoid metabolism. *Plant Cell*, 7: 1085–1097.

Ebel, J. and Costo, E.G., 1994. Elicitors of plants defence responses. *Int. Rev. Cytology,* 148: 1–36.

Facchini, P.J., 2001. Alkaloid biosynthesis in plants: Biochemistry, cell biology, molecular regulation and metabolic engineering applications. *Ann. Rev. Plant Physiol. Plant Mol. Biol.,* 52: 29–66.

Pushpangadan, P. and Govindarajan, R., 2003. Role of medicinal plants in nutraceutical sector for nutritional security and export promotion. In: *Proceeding of the Second International Conference on Herbal Drugs and Health Tourism,* UTTAHAN – Centre for Sustainable Development and Poverty Alleviation, Allahabad, pp. 234-242.

Pushpangadan, P. and Govindarajan, R., 2005. Need for developing protocol for collection/cultivation and quality parameters of medicinal plants for effective regulatory quality control of herbal drugs. In: *Proceedings – International Conference on Botanicals,* Kolkata, India, pp. 63-69.

Trapp, S. and Croteau, R., 2001. Defensive resin biosynthesis in conifers. *Ann. Rev. Plant. Physiol. Plant Mol. Biol.,* 52: 689–724.

Chapter 13

Strategies for Sustainable Chemical Processing

M.P. Sukumaran Nair

Special Secretary to Government of Kerala,
Chairman, Chemical Engineering Division, Board of the Institution of Engineers (India)
Director, Centre for Green Technology and Management
www.drmpsukumarannair.com

ABSTRACT

Environmental considerations assume significant importance in chemical and petro-chemical processing, which contribute to the generation of effluents and emissions capable of degrading environment. Effective environmental management addresses numerous issues relating to pollution control and ensuring safety and maintaining sustainability in this industry. In the recent years, several studies in different parts of the world have focused this issue with the objective of identifying key issues in environmental protection in relation to the hydrocarbon processing industries. In this paper the author discusses the status of environmental management, challenges, Green Manufacturing (GM), reduction of greenhouse gases, water intensity, key issues in environmental management in the hydrocarbon processing industries. The policy measures call for Annual Environmental Status Reports (AESR) along with the financial performance reports. The feedback on these reports from the concerned stakeholders may also be used for continued improvement of existing systems.

Keywords: *Hydrocarbon processing industries, Green manufacturing, Environmental management, Chemical engineering, Sustainable environment, Chemical effluents, Greenhouses gases, Clean technologies.*

Introduction

All around and exalted awareness of environmental degradation caused by industrial activities in the hydrocarbon processing sector on account of greenhouse gas emissions, stratospheric ozone depletion, acid rain and acidification, eutrophication, soil degradation, technological hazards, chemical mists and fog, all with potential damage to human beings is indeed a topic of much discussions at both national as well as international forums.

Environmental considerations, therefore assume significant importance in chemical and petro-chemical processing, which contribute to the generation of effluents and emissions capable of degrading environment and thus, industry operators are challenged on the one hand with the dire necessity of increasing production of goods and services needed for the people while on the other to effectively tackle the consequential environmental issues to avoid damages to the community. Effective environmental management perspective thus assumes paramount importance to address the numerous issues relating to pollution control and ensuring safety and maintaining sustainability in the industry. Clean development initiatives in chemical industries is a major development in this line.

The growth of industrial processing hitherto has been guided mostly by the necessity of increasing production at lower costs which led to serious environmental degradation of water resources, soil and air around the processing plants. Worldwide the focus of pollution control in the industry has shifted from end of pipe treatment to source reduction, avoiding pollution, clean technology and sustainable development. Hence, it has become imperative that environmental considerations shall play a substantive role in the future development of the industry especially at a time when more and more industrial activities are being undertaken in the developing countries.

In the recent years, several studies in different parts of the world focused this issue with the objective of identifying key issues in environmental protection in the different industrial processes, assessing as to what extent the national and international norms or guidelines regarding pollution control and environmental management are implemented in the industries, understanding the problems encountered in environmental management and exploring the reasons for the non-compliance. Suggestions were also made on the basis of the above studies to develop guidelines of an environmental policy and adopt cleaner technologies that will foster development with least degradation to the environment.

Status of Environmental Management

Today, most chemical processing units, be it in the developed or developing countries, have specific environmental policies and their emissions, effluents and waste disposal are guided by the stipulations of the regulatory authorities. New plants that are being built have modern technologies where considerable integration has taken place at the inception stage itself to see that pollution prevention is a part of the process design itself. Other units, which came to existence earlier, are now operated with additionally built state of art pollution control facilities. The pollution and environment control departments attached to the plants usually exhibit meticulous care to see that the above objective is achieved. Thus, now-a-days effective control facilities exist in most of the processing units and they are operated with due diligence. The stipulations of the pollution control and environmental protection agencies are also within the achievable limits of the available technology. Still excursions, at times, occur in the parameters on account of start up, shut down of plants or may be due to accidental situations.

Several national as well as international standards covering a wide range of parameters have been developed to specify the emissions and effluents from chemical industries. These include pH, ammonical nitrogen, nitrates, fluorides, phosphates, total suspended solids, oils and fats and chemical and biological oxygen demands in the effluent streams, particulate matter, nitrogen oxides, sulphur oxides and carbon monoxide in exhaust streams. Radioactivity, toxicity, presence of heavy metals, organics, biological pollutants and pathogens etc are also monitored in specific cases.

Clean Technologies

Current developments in environmental chemistry and chemical engineering have helped plant operators to reduce effluent generation at source and thus eliminate treatment and disposal itself. Very often treatment of pollutants emanating from industrial operations is linked to the processing technologies adopted in the plants. Over the years the specific consumption of raw material and energy for manufactured products has registered a continuously declining trend with the adoption of efficient technologies and best operating practices at the plant level. This invariably contributes to achieving better environmental standards through reduction in emissions, effluents and solid waste per tonne of product manufactured.

Further improvements towards better environmental quality may require major design changes involving additional investment or going for a newly proven and commercialized process. This is a costly option and hence efforts in this line are limited unless it brings about economic incentive by way of increased productivity, lowering of energy consumption etc.

In the case of products having high water intensity, there is an economic benefit in reusing treated effluents so that water conservation is achieved.

The start up and shut down of plants are situations that may lead to an increased level of pollution of the environment compared to its normal operation. Hence, most plants are equipped with specific provisions to take care of such situations.

Most of the pollution prevention methods implemented in the industries follow prescriptive approaches which follow standardized procedure built around questionnaires and check lists. The new approach is to adopt a more descriptive approach in which process operators are challenged to attack pollution problems and devise new and innovative ways for solving them.

Managements undertake substantial efforts to develop green belts and maintaining greenery around process plants to reduce the impact of green house gases. This is an important step in the direction of sustainable environmental control.

Establishment of ISO 14000 Environment Management Systems and a corporate environmental set up for regular monitoring and control is another major step in environmental protection. These systems are intended for continuous improvement of existing operations from the environmental angle. Certain industries have adopted zero effluent approach incorporating total recycle and reuse of effluents back to process, though it still remains more a concept than its effective implementation to a reasonable degree of reliability.

European process plant operators use the Best Available Techniques (BAT) in their plants for environmental control. Both effluent specific standards and product specific standards are available.

In India and many other developing countries systems are employed to the extent of controlling and reducing pollution from plants with in the limits set by the statutory authorities, *i.e.;* the Pollution Control Boards (PCB). Operating units do not put in further efforts for reducing the pollution effects beyond the limits prescribed by the Pollution Control Boards (PCB). This is primarily due to lack of incentives to encourage additional investment towards improved technology to go for better environmental quality.

Most operations emit large quantities of carbon dioxide (CO_2), which is a major green house gas to the atmosphere. There are no emission standards for carbon dioxide as prescribed by the statutory bodies. Attempt to reduce green house gas emissions all over the globe to tackle climate change may bring in specific limits for carbon dioxide emissions or call for effective measures for sequestration in future.

Every processing unit imposes certain environmental burden to the local environment and its impact categories are acidity, global warming, human health effects, ozone depletion, photochemical smog, aquatic oxygen demand and eco-toxicity to aquatic life etc. A parametric assessment of the contribution of each of these components can be used to compare yearly performances of plants.

The necessity of maintaining a safe work environment for employees and the neighboring community is well recognized. For this purpose extensive hazard and risk analysis using techniques such as Hazard Operability (HAZOP) Studies and Quantitative Risk Assessment (QRA) are conducted based on which safe systems, work practices and risk reduction measure are adopted. The environment management plans of the production units are capable of mitigating the risk from most expected crisis situations barring those from nightmare incidents such as earthquakes, sabotage etc.

Information to the public regarding the environmental consequences of these plants is very important. The communities associated with these units have a right to know the environmental risk they are subjected to.

In most countries it is now mandatory that an Environment Impact Assessment (EIA) be done prior to implementation of a project having large scale environmental consequences. A proper Environment Management Plan (EMP) is also to be in place before the unit start operating.

Environmental Challenges

Climate change across the world, depletion of ozone layer in the outer atmosphere, loss of biodiversity elements such as migratory species and important genetic resources, widespread degradation of land, urban air, forests and natural waters and marine ecosystems, accumulation of persistent organic pollutants in nature are major global environmental concerns. These issues have an impact that transcendent national boundaries and hence require global solutions. We have over 200 international legislation governing environmental issues and together with currently available technology and adopting best practices mitigation of further degradation is possible. India for example, has adopted a national action plan for climate change and framed a comprehensive auto fuel policy that consider among other things, availability and security of supplies, vehicle technology, cost effective emission reduction, fiscal measures and institutional means to bring about progressive improvements by reducing vehicular emissions on ambient air. The fuel cell as a power source is becoming a viable alternative to the internal combustion engines with least environmental impacts. Thus concerted efforts are required both at the national and international level to stop further degradation and undo the damages already done.

Stress on Environmental Health

National environment policies shall foster efforts for sustaining environmental health of the people and shall call for a discrete assessment of pollutants entering the natural environment from human interventions in terms of their toxicity, persistence, mobility, bioaccumulation and methods available for source reduction and control mechanisms.

Shift Towards Alternatives

More and more industries are switching over to environment friendly raw material and energy resources in order to improve own sustainability. Several examples are available.

The use of natural gas (predominantly methane, CH_4) as a relatively benign raw material and energy source than other petroleum feedstock as naphtha and fuel oil and coal to produce ammonia which is the basic building block of the nitrogenous fertilizer industry indicate a shift to reduce pollution problem. This route has reduced CO_2 emissions, low waste generation and low energy intensity for the product.

The use of water in place of organic solvents to replace volatile organic compound (VOCs) as reaction medium in organic synthesis is another important innovation. Use of ethylene dichloride (EDC) for extraction of food items have been replaced by *n*-Hexane.

Manufacture of sulphuric acid from pyrite roasting has a lesser impact than using elemental sulphur. Recycling of metals, recovery of metals from spent catalysts, sludge from metallurgical operations etc are intended for reducing the impact of large scale mining of metals and minerals.

Begasse from sugar industry is used as a raw material for paper industry in place of wood pulp; treated softwood, coir and bamboo composites are extensively used in place of hardwood for furniture. Several oxidation reactions involving air are being replaced with gaseous oxygen to ward off formation of toxic nitrogen oxides. Use of new generation polymers and plastics in place of metal in highly corrosive applications etc are intended to reduce the environmental burden arising out of products manufactured and lead to sustainability.

The construction industry is being increasingly encouraged to use locally available materials to replace steel, glass, wood, and cement.

Use of hydrogen in place of fossil fuels in engines is in advanced research stage and once it becomes technically and economically feasible, offer an environment friendly option for transportation. Natural zeolites find extensive applications in place of alkyl benzene sulfonates (ABS) as detergents. Power generation using vacuum residue from crude refining operations is a sustainable option to dispose of the unmarketable end product. These bottoms after improved crude utilization process such as hydro cracking etc are highly viscous, heavy, difficult to handle and as such are not marketable. The quantity of residue from refineries varies from 20to 40 per cent of the throughput depending on the crude characteristics and secondary process employed. Refiners are looking for alternative residue utilization strategies for producing lighter, high-value products to help them remain competitive in a market where the demand for light oil products is steadily increasing.

Several biofuels are being used in the automotive industry. Ethanol and bio-diesel have emerged as reasonable alternatives to the conventional petroleum fuels. These are of agricultural origin and can be easily blended with hydrocarbon fuels unto 50 per cent blends and can be used in existing vehicles with out any modifications. Biofuels are renewable resources, non-toxic, bio-degradable and have reduced inflammability and their performance is also superior. Ethanol blended gasoline have become a commonplace fuel in many countries. Ethanol largely available as a byproduct of the sugar industry is non-toxic and is thus environment friendly causing no harm to soil, water bodies and public health. Being an oxygenated fuel ethanol enhances the combustion of gasoline and effectively lowers emission rates from engines.

The use of natural fibbers like coir, jute etc put least burden on the environment vis-à-vis other products for similar applications such as nylon, polyester, poly ethylene etc. Coir geotextiles find extensive application in preventing soil erosion and landslides.

The use of biotechnology products reduce the intensity of normal cropping in terms of inputs like fertilizer, water and have better resistance to adversities like pest infection, crop losses and increased productivity. Several BT products help to reduce environmental footprints.

Recycling of paper and use of natural polymers in place of plastics are greener options. Currently, 150 million tons of plastics are produced a year from fossil fuels all around the world. Plastic products have gained universal use not only in food, clothing and shelter, but also in the transportation, construction, medical, entertainment etc. There is a growing demand for biodegradable plastics as a solution to global environmental and waste management problems.

Research on biodegradable plastics and polymers has been carried out worldwide with the aim of achieving a balance between human activities and the natural environments Ideal biodegradable plastics are defined as materials, which are completely degraded to CO_2 and H_2O by the action of microorganisms. The resources of biodegradable plastics and polymers are mainly classified into biosynthetic polymers such as poly hydroxyalkanoic acid, plant polysaccharides such as cellulose, starch and xanthan, and chemical synthetic polymers such as poly lactic acid, poly ε-caprolactone and poly aspartic acid. Biodegradable plastics are expected to replace plastics derived from petroleum products in natural environment such as agricultural and fisheries, civil engineering and construction materials and toys where recovery and reuse are difficult and where composting of organic waste is effective such as food packaging, hygienic products on account of its specific features, such as slow release, water retention, medical use, low oxygen permeability and low melting temperature.

Green Manufacturing

Green Manufacturing (GM) aims to prevent pollution and save material and energy through innovation and development of new knowledge that reduces or eliminates environmental damages right from the design, manufacture, and application of products or processes. Apart from use of benign materials, changing technologies for processing also add on to the environment friendliness of manufacturing industries. Existing processes also undergo sea changes to become environment friendly. Thus GM requires rethinking of manufacturing systems by pursuing environmentally related goals and objectives, non-traditional manufacturing processes, new marketing strategies, and product design based on a life-cycle approach.

Improved catalysts increase conversion and yields, reduce recirculation and increase outputs. The best illustration is the innovation is the development of ruthenium catalyst in place of conventional iron catalyst for the ammonia reaction which helped to improve conversion threefold, reduce size of plant equipment and rendered higher plant capacities viable.

In sito generation and consumption of hazardous and toxic intermediates and thus avoiding storage and transport is another option. An example is the manufacture of methyl iso-cyanate and its immediate conversion to pesticides without going for storages, a lesson we have learned form Bhopal.

Other sustainable improvements in reaction engineering include carbonylation using dimethyl carbonate instead of using phosgene or carbon mono oxide, use of solid catalysts to minimize waste, attempt to manufacture petrochemicals from renewable sources, produce alpha olefins from fatty acids rather than petroleum products, avoiding use of toxic acids, catalysts, solvents in all applications and going for photo catalyzed reactions using natural dyes as catalysts instead of heavy metals etc.

Use of low analysis fertilizers with nutrient content sufficient only to meet the demand of the plants during the cropping season minimizes leaching to the environment.

The development of membrane cell process for caustic soda in place of mercury cells is an example. This has helped the industry to ward off mercury contamination and the byproduct hydrogen could be used for food grade applications.

Membrane processes and pressure swing adsorption have come of age for physical separation of gases in place of chemical absorption and regeneration.

The harshness of chemical reaction as depicted by elevated temperature, increased concentration, high pressures, large reactor volumes, corrosion tendencies, flammability characteristics etc are being considerably brought down through technological innovation. Now-a-days several reactions are being carried out under lower temperature, pressure and concentration with improved catalytic efficiencies. The best example is again from the fertilizer industry where ammonia was used to be synthesized 350 atmospheres pressure two decades ago, has been lowered down to operate at as low as 80 atmospheres. The potential for recycling and reuse within are being exploited considerably and modern plants are built with such integrated facilities.

Promote Integrated Complexes

Refinery, fertilizer, power and petro-chemicals are themselves major investment and high technology decisions and very often these units are put up by different agencies and function as independent companies. Technology brings in lot of scope for exploiting the synergy with in these units, which could play a major role in improving the bottom line of current operations of these units. Integration of refineries, fertilizer and petro-chemical plants, and power generation units at the planning phase it self to develop integrated complexes will help to drastically reduce emission and other pollutants and ensure optimized operation.

Review Existing Control Limits for Pollutants

The present standards for discharge of effluents from industrial units are technological limits attainable through application of available technologies for abatement and control available at the time of specifying these norms and are not based on their long-term health effects. Revised standards based on the health impacts of each of the pollutants may be developed incorporating the advancement in this area. While doing so care must be taken to see that the prescribed limits are technologically achievable and are within the means and reach of the industry.

In process plants, normally systems are employed to the extent of controlling and reducing pollution from plants with in the limits set by the statutory authorities. The units do not put in further efforts for reducing the pollution effects beyond the limits prescribed by the Central Pollution Control Board (CPCB) in the interest of public health. This is primarily due to lack of incentives to encourage additional investment towards improved technology. Hence, industrial units shall be encouraged to go beyond compliance and become more environment friendly.

Harness Environmental Biotechnology

Environmental biotechnology employ living organisms –flora and fauna-engineered to exhibit specific traits in order to identify, control or prevent pollution. This technology has been applied to clean up hazardous waste sites more efficiently than conventional methods, thereby reducing the need for incineration or extraction-based methodologies. Bio-remediation has been applied to the cleanup of numerous varieties of pollutants, including heavy metals, persistent organic pollutants, explosives, sewage and industrial waste. Because of the prevalence of tropical climate, biological processes for pollution control have an edge over chemical processes and are more efficient. Modern developments such as recombinant and genetically engineered organisms find extensive application in biological processes for pollution control and bioremediation.

Reduce of Emission of Greenhouse Gases

Encouragement through adequate financial incentives shall be made available to those intending for voluntary reduction of greenhouse gases and those resulting in climate change etc. The extension of natural gas pipelines, harnessing clean coal technologies, gasification of biomass, and development of hydel and nuclear sources for power generation shall be encouraged.

Reduce Water Intensity

Chemical processing is a highly water intensive industry. Availability of good quality water for the community and industry is going to be a major problem in the coming years. In order to address the issue of availability of adequate water for industries recycling and reuse to the extent possible may be resorted to. It has become a boon that membrane technologies have come of age and cost effective these days.

Develop Pollution Inventory Database

The policy shall strive to develop a national level pollution inventory database and ensure that pollutants get reduced over a period of time during the development process.

Changing Role of the Regulator

Environmental regulatory authorities in developing countries may be encouraged to become solution providers to the industry rather than being mere policing agents.

Hazardous Waste Disposal

Management of hazardous waste materials generated in the industries has become a major concern of plant operators from the environmental angle. Hazardous waste may be a solid, semi-solid or non-aqueous liquid which because of its quantity, concentration or characteristics in terms of physical, chemical, infectious quality capable of significantly contributing to an increase in mortality or an irreversible damage. Left uncared or improperly treated, stored, transported and disposed, they are capable of posing a potential hazard to human health and neighboring environment. A waste material is classified as hazardous if it exhibits whether alone or in contact with other wastes or substances, any of the characteristics such as corrosively, reactivity, ignitability, toxicity, acute toxicity or infectious property. These substances either created as by-products of the industry or as residues of the process adopted are highly toxic and are capable of causing irreversible damages to the environment. Most industries have identified such materials coming under the purview of hazardous wastes from their operations and are subsequently classified.

A lot of hazardous waste is generated in-house as a result of several industrial operations and there are imports too for recovery of valuables etc. Determined efforts from the part of local Governments to put up state of art chemical hazardous treatment and disposal facilities (CHTDF) have become necessary.

Emergency Planning for Disaster Mitigation

Local level emergency planning for disaster preparedness in case of natural calamities and man made disasters is important. Effective mechanisms for mitigation of hazards shall be developed under the district administration. The programme may be coordinated on the lines of the Awareness and Preparedness for Emergencies at Local Level (APELL) project of the United Nations Environment Programme (UNEP).

Key Issues

Thus, the key issues in environmental management in the hydrocarbon processing industries are identified as pollution from solid waste resulting in contamination of land space, liquid effluents endangering water streams and ground water resources and gaseous emissions degrading the quality of atmospheric air, risk to life from operational incidents to people and property in the industry and those in the neighborhood of these units due to storage, handling, transport and use of large quantities of inflammable and hazardous chemicals and hydrocarbons, large scale depletion of natural resources, raw materials, energy resources and water and contribution to global warming due to emission of greenhouse gases.

Studies also reveal that the units have been successful in controlling pollution from their operations to the level prescribed by the statutory authorities and as required by the law. The best available technology for pollution control and environmental management are being used and it compares well with such practices being adopted internationally.

Generally there is a good deal of compliance by all units to the standards prescribed for discharges of effluents. Often units are committed to attain the norms for various parameters as stipulated by the Central Pollution Control Board. Units even go to levels pollution control beyond compliance if there are sufficient economic incentives for making the required additional investments. In other situations no attempt is made by units to achieve better control of pollution beyond the statutory limits.

Addressing Challenges: National Environment Policy

The important problems encountered in environmental management in the CPI are lack of:

1. Incentive for continuous improvement in the direction of pollution reduction beyond the compliance limits of the Central Pollution Control Board,
2. Integration of environmental concerns in to the core of the business strategy and
3. Sufficient transparency with regard to environmental information.

To effectively address the above problems and foster development an apex level environmental policy incorporating clean development strategies with the following elements will be required.

Role of Top Management

The first and foremost guiding principle of an environmental policy facilitating growth of the industry is to ensure the unstinted commitment, involvement and action oriented approach of the top management of the organization in achieving the set environmental goals.

Environmental Policy Statement

Top management shall codify their environmental commitment, values and perceptions in a documented policy. The policy shall be relevant to its activities, products, and services and taking into account of its implications on the different stakeholders. Attempts for improving energy efficiency, resource productivity and use of renewable source of energy and raw materials need special mention in the policy.

Environment, Health, Safety (EHS) Vision Statement

Every unit shall have an environment, health and safety a vision statement depending upon the nature and scale of its operation and specifying its current thinking and aspirations of the future. They shall adopt a national pollution prevention policy that encourages source reduction and environmentally sound recycling as a first option, but that also recognizes safe treatment, storage and disposal practices as important components of an overall environment protection strategy.

Environmental Targets

The environmental targets *i.e.*, the qualitative and quantitative changes that are to be brought about to bring in more environment friendliness in the industry and acceptance to the community around are to be clearly set. Steps that are envisaged for minimizing environmental impacts, reducing emissions of toxic gases and those causing global warming and improving the current levels of employee health, safety and pollution prevention are to be specified. The target must also address achieving zero accidents at work places, reducing incidents of work related diseases and overall reduction of the risk exposure to the employees as well as the community around. It focuses achieving sustainable development and eco-efficiency as a new business perspective for the industry through production and innovation integrated environmental protection, responsible product stewardship and aiming total quality improvements. It is desirable that the environmental friendliness of the industry shall improve year by year through implementation a guided approach and action plan. For this purpose existing environmental burden imposed by these units have to be quantified by considering suitable indices for every environmental aspect.

Control Strategies

The policy shall provide for the use of legal, financial and social instruments, which influence the behavior of companies, citizens, public bodies and authorities for achieving the objectives of the policy. Existing and innovative control mechanisms such as statutory provisions, stipulations of the various regulatory bodies may be used. Industry may be asked to go for the currently best available technology for pollution abatement. During the interim phase strategy of monitoring comparison with set standards and penal action wherever required shall continue. Plants shall be operated to standards that will comply with the requirements of appropriate national and international legislation and codes of practice. The government shall formulate country specific Best Available Techniques (BATs) for the industry to facilitate continuous improvement in environmental management. Technically and economically feasible regulatory as well as non-regulatory measures are also suggested to improve environmental management in chemical processing operations. Fiscal incentives may be provided to encourage adoption of technologies that reduce pollution.

Risk Management

It is necessary that the management shall ensure that potential health, safety, and environmental risks associated with the activities are assessed early to minimize and manage adverse effects and to identify opportunities for improvement. A workable Disaster Preparedness and Emergency Management Plan (DPEMP) shall be kept ready to mitigate any such situations in the unlikely event of its occurrence.

Staff Training

Necessary and state of art training may be given to the concerned people responsible for environmental management. This should include keeping them abreast of the new developments, technologies and practical tools, accident investigation, environmental impact prediction, selecting appropriate protective equipment, implementing emergency response plans as and when necessary and so on. They may be trained to learn from previous incidents and similar experiences. They must be made conversant in the corporate environmental management systems and the proposed action plan for its implementation. In short necessary capabilities must be available in-house with all organizations to tackle probable emergency situations that are likely to arise.

Monitoring

The policy shall call for regular and meticulous environmental performance monitoring to keep track of the environmental burden imposed by the company and watch the direction of its progressing trends. Quantitative as well as qualitative approaches may be used for this purpose. Emissions, waste streams, hazardous waste, disturbance, resource depletion etc shall be addressed accordingly. Commitments towards targets for Responsible Care and Social responsibility may also have to be assessed.

The current operations shall be regularly and systematically assessed for the purpose both of identifying and correcting any element which may put human beings, real property or the natural environment at risk of nuisance or damage and of establishing a basis of safety-related improvements of processes and products. Any new process and product as well as any new information of existing processes and products should be thoroughly analyzed with regard of their health, safety and environmental implications.

The concerned authorities are kept well informed of the operations and of their health, safety and environmental implications. Any incident entailing a risk of environmental disturbances or of conflict with existing regulations should be promptly reported to proper authorities.

Public Information

Necessary provision may be made for sharing information on environment safety and health with the public is to be incorporated in the policy. The 'community right to know' Act or the 'right to information' Act in several states are intended for this. The policy shall provide for involvement of the community and working with active environmental groups in the region in bettering the environmental situation and thereby enhancing public perception of the industry.

Annual Reports

The policy shall call for Annual Environmental Status Reports (AESR) along with the financial performance reports. Such reports are now available from many operators around the world. The feedback on these reports from the concerned stakeholders may be used for continued improvement of existing systems. The policy document shall be integrated with the national environmental plan of the country.

References

Amory, B., Lovins, L., Hunter Lovins and Paul Hawken, 1999. A road map for natural capitalism. *Harward Business Review*, May-June.

Dyer, James A. and Mulholland, Kenneth L., 1998. Follow this path to pollution prevention. *Chemical Engineering Progress*, American Institute of Chemical Engineers (AIChE), New York.

Environmental Impact Assessment–A Manual, Ministry of Environment and Forests, Government of India, 2000.

Lemkowitz, S.M., Korevaar, G., Marmsen, G.J. and Pasman, H.J., 2001. Sustainability as the ultimate form of loss prevention: Implications for process design and education. In: *Proceedings of the 10th International Symposium on Loss, Prevention and Safety Promotion in the Process Industries*, European Federation of Chemical Engineering, Stockholm, Sweden.

Nair, Sukumaran M.P., 2004. Integrating safety and environmental concerns in the development of chemical processing industries. *Loss Prevention*, Praha, Czechoslovakia.

Risk Analysis in Process Industries, Report of the International Study Group set up by the European Federation of Chemical Engineering, Institution of Chemical Engineers, UK, 1990.

Volodin, N. Igor, 1997. *Pollution Prevention and Pollution Control*. UNIDO/UNDP Initiative for Clean Production Centers.

Chapter 14

The Impact of Educational Tourism on Sustainable Economic Development

***Mohammad Hossein Karim*[1] *& Malihe Esmaeili*[2]**

[1]*Counselor and Director, Science and Education,*
Embassy of Islamic Republic of Iran in India
Associate Professor, University of Economic Science, Tehran/Iran
E-mail: irda87@gmail.com
[2]*Ph.D. Scholar, University of Pune*
E-mail: esmaeili06@yahoo.com

ABSTRACT

Education and tourism industries are growing very fast in recent decades and are leading to growth of these industries from an economic and social prospect.

Sustainable development stands in acquiring a relevant measurement which is a balance between the development of tourism and tourist regions from one side and protecting and preserving the environment, resources and values of local communities by mass preservation training, on the other side. This also argued that supporting the view that sustainable tourism can be achieved only as an integral part of sustainable regional development based on conservation and education.

Edu-tourism contributes to sustainable economics and is an alternative tourism that incorporates and goes beyond aspects of business tourism, sport tourism, heritage tourism, eco-tourism, community tourism and other types of tourisms. It is a hybrid that takes into account the visitor's need for enjoyment in a stress-free environment, and is grounded in the local people's deep cultural roots, social and economic needs, and active participation.

This paper focuses on educational tourism as one of the strategies which improves tourism development. It requires coordinated, sustained and organized methods and procedures to achieve its social and economic potentials. The paper explores the impact of edu-tourism on countries' revenue generation and sustainable economic development with a special reference to edu-tourism in USA, UK, Canada, and Australia as developed countries, India and Malaysia as developing countries and the utilization of the findings in development of edu-tourism in Iran with the moto that 'knowledge is power'.

Keywords: *Sustainable development, Educational tourism, Economic development.*

Introduction to Tourism Industry

Tourism means employment and wealth to many nations across the world. There is a promised and sustainable growth in this sector with over quarter billion people working directly in this sector. It is predicted that tourism will be the highest money-spinner in the new century that has prompted many governments to embark special attention to this sector (Prasad and Sundari, 2009). According to Harris *et al.* (2002), Sustainable tourism is tourism that is developed and maintained in a manner, and at such a scale, that it remains economically viable over an indefinite period and does not undermine the physical and human environment that sustains and nurtures it. Technology is playing a big role in the changing dynamics of travel and tourism. Internet has enabled worldwide reach for host facilitators. Economists feel that tourism industry contributes substantially to the development of backward areas where smoke-generating industries have not been set up (Kulkarni, 2005).

The economical benefits of tourism are:

1. *Employment generation*: Tourism provides direct and indirect employment for millions of people.
2. *Development of employee skill*: Tourism is customer centric service and demands high skills of human interaction thereby it contributes to the skill-development of the employees.
3. *Foreign currency*: Tourism contributes significantly to the foreign currency needs of an economy.
4. *Contribution to the GDP (Gross Domestic Product)*: The contribution from tourism to GDP is around 11 percent in developed countries and around three percent in emerging economies.
5. *Multiplier effect*: This effect is very strong in tourism sector; The money received from tourism gets transferred from hand to hand and as its transfers, it creates a cycle of new income and expenditure; the higher the multiplier effect, the higher is the benefit to an economy.
6. *Growth of infrastructure*: Tourism needs well-structured roads, airports, seaports and etc. Development in tourism invariably develops infrastructure in a country and also helps develop the existing facilities and thereby create value to the region and local community.
7. *Balanced growth*: Tourism helps sustain regional development of an economy.

8. *Cultural exchange*: It leads to improve social, culture and entertainment diversification.

The growth of leisure time in combination to the increase of income and a desire for holidays has led to the growth of tourism. According to the World Tourism Organization (WTO), tourism is predicted to increase with future tourist arrivals growing to 1.6 billion by 2020 at an average growth rate of 4.3 per cent. Despite the effect of external factors such as the Asian Economic crisis in the late 1990s and the September 11 incident in 2001, tourism growth appears to be assured.

The growth of tourism has spread geographically with the market share of tourist arrivals reducing in Europe and increasing in Asia-Pacific area, which has the fastest growth rate of world tourist arrivals. There has also been a growth in tourism to developing countries with their share of tourist arrivals and expenditure increasing especially in destinations such as Eastern Europe, South America, the Middle East and Africa (Conrady, 2010). However, the growth of tourism on a worldwide scale, coupled with the search for new destinations and experiences, has added to the questioning of tourism impacts and calls for more sustainable types of tourism. Since 1980s the tourist marketplace has become increasingly specialized and segmented resulting in the growth of the niche markets such as rural tourism, ecotourism, adventure tourism and cultural heritage tourism. Furthermore, a number of educational and learning experiences within tourism appear to be increasing, while travel and tourism experiences especially for study or learning also appear to be increasing, illustrating the potential for educational forms of tourism (Brent *et al.*, 2003).

Sustainable tourism is: "Tourism that meets the needs of present tourists and host regions while protecting and enhancing opportunities for the future" (Misra and Sadul, 2008). Under the overall framework of strengthening national capabilities and promoting regional cooperation in sustainable tourism development, the plan proposes action at the national level and supporting action at the regional level. In this context, sis theme areas have been identified in which specific action is required. The theme areas are: human resources development in the tourism sector, the economic impact of tourism, environmental management of tourism, infrastructure development and investment for the tourism sector, facilitation of travel and regional and sub-regional cooperation in tourism development (Kandari *et al.*, 2004).

Tourism Industry: Facts and Figures

According to World Tourism Organization (WTO), tourism contribution to worldwide gross domestic product (GDP) is estimated at some 5 per cent which ranges from 2 per cent for countries where tourism is a comparatively small sector, to over 10 per cent for countries where tourism is an important pillar of the economy; the international tourism receipts reached to 852 billion US$ in 2009 and the overall export income generated by inbound tourism, including passengers transport, exceeded 1 trillion US$ in 2009. Tourism exports account for 30 per cent of the world's exports of commercial services and 6 per cent of overall export category; tourism ranks fourth after fuel, chemicals and automotive products. For many developing countries it is one of the main sources of foreign exchange income and the number one export category, creating much needed employment and opportunities for development. The contribution of tourism to economic activity worldwide is estimated at some 5 per cent and tourism's contribution to employment tends to be slightly higher relatively and is estimated in the order of 6-7 per cent of the overall number of jobs worldwide (direct and indirect).

According to World Travel and Tourism Council (WTTC, 2001) tourism currently generates 6 per cent of global Gross National Product and employs one in 15 workers worldwide. It is predicted that by 2011 it will directly and indirectly support one in 11.2 workers and contributes 9 per cent of Gross National Product worldwide (WTTC, 2001). Although the evolution of tourism in the last few years has been irregular, UNWTO maintains its long-term forecast for the moment. The underlying structural trends of the forecast are believed not to have significantly changed. Experience shows that in the short term, periods of faster growth (1995, 1996, 2000) alternate with periods of slow growth (2001 to 2003). While the pace of growth till 2000 actually exceeded the *Tourism 2020 Vision* forecast, it is generally expected that the current slowdown will be compensated in the medium to long term. (Figure 14.1)

UNWTO's *Tourism 2020 Vision* forecasts that international arrivals are expected to reach nearly 1.6 billion by the year 2020. Of these worldwide arrivals in 2020, 1.2 billion will be intraregional and 378 million will be long-haul travelers. The total tourist arrivals by region shows that by 2020 the top three receiving regions will be Europe (717 million tourists), East Asia and the Pacific (397 million) and the Americas (282 million), followed by Africa, the Middle East and South Asia. East Asia and the Pacific, Asia, the Middle East and Africa are forecasted to record growth at rates of over 5 per cent year, compared to the world average of 4.1 per cent. The more mature

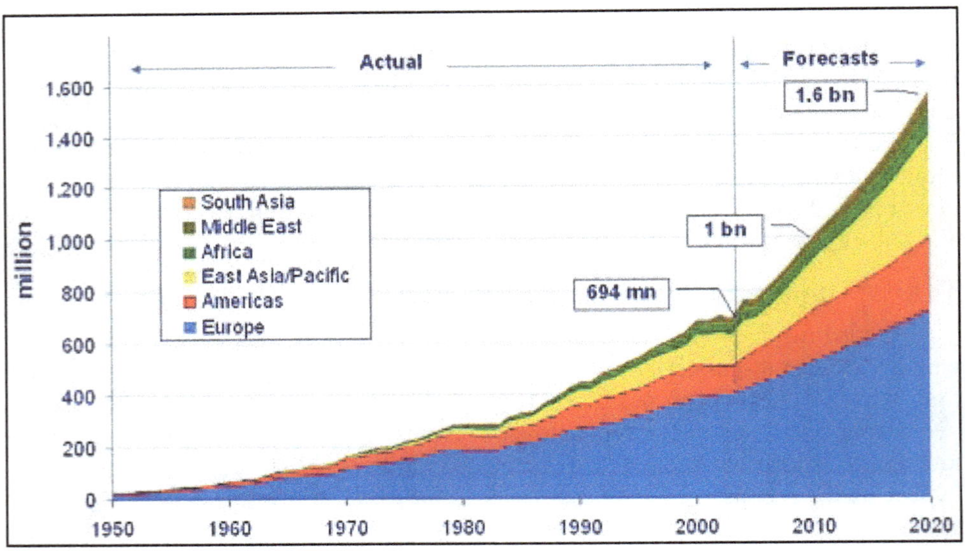

Figure 14.1: Tourism 2020 Vision/International Arrivals (*Source*: United Nations World Trade Organization)

regions Europe and America are anticipated to show lower than average growth rates. Europe will maintain the highest share of world arrivals, although there will be a decline from 60 per cent in 1995 to 46 per cent in 2020.

Tourism and the World Economy

International tourism receipts represented in 2003 approximately 6 per cent of worldwide exports of goods and services (as expressed in US$). When considering service exports exclusively, the share of tourism exports increases to nearly 30 per cent (Table 14.1). Tourism demand depends above all strongly on the economic conditions in major generating markets. When economies grow, levels of disposable income will usually also rise. A relatively large part of discretionary income will typically be spent on tourism, in particular in the case of emerging economies.

Table 14.1: World Export of Merchandise and Commercial Services (Balance of Payments, Goods and Services Credit) 2003

	US$ Billion	*Share (Per cent)*	*Share (Per cent)*
Total		**9,089**	**100**
Merchandise exports	7,294	80	
Agricultural products	674	7	
Mining products	960	11	
Manufacturers	5,437	60	
Other	223	2	
Commercial services	1,795	20	100
Transportation	405	4	23
Travel	525	6	29
Other	865	10	48

Source: World Trade Organization, World Tourism Organization.

A tightening of the economic situation on the other hand, will often result in a decrease or trading down of tourism spending. In general, the growth of international tourism arrivals significantly outpaces growth of economic output as measured in Gross Domestic Product (GDP). In years when world economic growth exceeds 4 per cent, the growth of tourism volume tends to be higher. When GDP growth falls below 2 per cent, tourism growth tends to be even lower. In and GDP at 3.5 per cent, *i.e.* tourism grew on average 1.3 times faster than GDP. In the period 1975-2000 tourism increased at an average rate of 4.6 per cent a year.

Educational Tourism

Understanding Education and Learning

Education has been defined as 'the organized, systematic effort to foster learning, to establish the conditions, and to provide the activities through which learning can occur' (Smith, 1982).

The higher education market is expanding rapidly as income levels rise, thus allowing for greater expenditure on education. The growing potential market for the travel and tourism industry amongst schools, universities and further educational institutions is one that should not be ignored and needs to be better understood by the tourism industry (Brent *et al.*, 2003).

What is Education Tourism?

The term education tourism or edu-tourism refers to any "program in which participants travel to a location as a group with the primary purpose of engaging in a learning experience directly related to the location" (Bodger, 1998).

Education and learning has more recently become an increasingly important and recognized component of travel activity and travel experience with an increasing amount of tourist activity involving some form of either formal or informal education and/or learning.

Edu-tourism is an alternative tourism that incorporates and goes beyond aspects of business tourism, sport tourism, heritage tourism, eco-tourism, and community tourism. It is a hybrid that takes into account the visitor's need for enjoyment in a stress-free environment, and is grounded in the local people's deep cultural roots, social and economic needs, and active participation. Edu-tourism creates reciprocal partnerships, through which all parties learn and contribute their respective resources, expertise, knowledge and skills particularly in higher education level. It involves host communities in a dynamic and self-defined form of economic and community development. The result is something that makes the whole experience qualitatively different from traditional leisure tourism.

Edu-Tourism benefits the geographic region as it:

- ☆ Improve sustainable development by offering an additional sustainable source of tourism revenue in the geographic region,
- ☆ Increase growth through further opportunity to showcase, develop and incorporate local culture and international knowledge.
- ☆ Provide respect and honours the local people/businesses and involves them in their geographic region in a more cohesive manner.
- ☆ Benefits distribution, the benefits will be broadly supplied among residents, tourism enterprises, governments, citizens, visitors, educational institutions, etc.

This fast growing sector provides a creative way to combine and promote a region's education/learning opportunities and unique tourism strengths, creating social and economic harmony for its people resulting in a more sustainable tourism product, on the bases of education tourism and long term tourism policy.

With the economic and demographic rapid changes and the many shifts within the travel industry, the need for more creative and sustainable approaches to market destinations and to fulfill the traveler is increasing.

Educational Tourism in Developed Countries

Edu-Tourism in United States

International educators, however, point out that the United States remains the top destination for foreign college students, despite efforts by colleges in Australia, Canada, and elsewhere to attract them. Some 671,616 international students attended U.S. institutions in 2008-9 (Table 14.2), an increase of almost 8 percent from a year earlier. First-time-student enrollments grew even more robustly, by nearly 16 percent. But the rosy data highlighted in the annual "Open Doors" report may obscure some potentially worrisome trends. Though graduate programs typically rely more on international students, enrollment grew far more strongly at the undergraduate level, where the number of students jumped 11 per cent, than at the graduate level, where enrollments climbed a little more than 2 per cent. What's more, the increase in students pursuing undergraduate studies was largely dependent on enrollment from China, which shot up by 60 per cent.

"There's still more talent out there than can be accommodated," says Peggy Blumenthal, executive vice president of the Institute of International Education, who notes that international students contribute an estimated $17.6-billion to the U.S. economy. In 2008-9, there were 269,874 international undergraduate students and 283,329 graduate students. (Students enrolled in intensive-english or other nondegree programs and recent graduates who are temporarily staying in the United States to work through the Optional Practical Training program are also counted in overall figures.)

India and China, which predominantly send graduate students, rose to become the top two suppliers of international students to American colleges. And the September 11, 2001, terrorist attacks convinced some parents that the United States was not a safe place for their teenage sons or daughters to study. This time around, all eyes are on India. It and China account for more than 45 per cent of all foreign students enrolled in American graduate schools. The rate of enrollment growth of Indian graduate students studying in the United States has slowed significantly, from 14.4 per cent in 2007 to just 4.3 per cent in 2008.

Key source countries like China and India simply cannot meet domestic demand for higher education, and a rising middle class in those nations means more families have the money to send their children overseas to study. Despite discouraging signs this year, Indian enrollments, including both graduate and undergraduate students, actually posted a healthy 9.2 per cent increase in 2008, according to Open Doors. Other countries saw even larger gains. Enrollments from Vietnam climbed 46.2 per cent, matching the previous year's increase, while enrollments from Saudi Arabia, fueled by a large government scholarship program, jumped 28.2 per cent. But the most significant development was the growth in Chinese students, whose numbers swelled by 21.1 per cent over all and by nearly 60 per cent at the undergraduate level. What's behind that trend? Thanks to a push by their government to make secondary education universal, more Chinese students are seeking college degrees, but there are not enough colleges, and too few high-quality institutions, to meet the need. A decline in the value of the dollar has put an American education in reach of middle-class

Chinese families–who probably had already been salting away much of their disposable income to pay for education.

Table 14.2: Foreign Students in U.S. by Region of Origin, 2008-9

Region	*Per cent of Foreign Students*
Asia	62 per cent
Europe	13 per cent
Latin America	10 per cent
Africa	5 per cent
Middle East	4 per cent
North America	4 per cent
Oceania	1 per cent
Total	671,616

Note: Percentages are rounded and so do not add to 100 percent.

Source: Institute of International Education, USA

As shown in Table 14.2, 62 per cent of foreign students in US are from Asia, where the population is high, the income level of people is increasing and countries are developing so the demand for higher education is high but the number of universities and educational institutions are not enough to meet this huge demand, so Asian students prefer to go abroad especially to US in order to get educated; 13 per cent of foreign students of US are from Europe, where the population is much lower than Asia, people have much higher income level comparing to Asians but the demand for educating out side Europe is low due to the existence of high number of universities and educational institutions which themselves are trying to attract foreign students from rest of the world; 10 per cent of foreign students of US are from Latin America where the demand for higher education is high but people have lower income level comparing to Asians in order to be able to over come the expenses of sending their children to other countries for higher education; Africans count 5 per cent of foreign students of US, the reason could be that people are very poor and can not afford education and those who can afford, prefer to study where the cost of living is low such as India or other Asian countries.

According to the Institute for International Education (IIE), for the 1993-94 school year there are more than 575,500 foreign students American colleges and universities. But, in reality, foreign students significantly exceed the 4.6 percent of all student enrollment, because the IIE data miss foreign students in the country illegally and others in the country on some different type of visa. In addition, the IIE data do not include foreign students in secondary schools. The Council on Standards for International Educational Travel puts the number of high school foreign exchange students at 27,742 in the 2002-03 school year. Figure 14.2 shows the increase in the number of foreign students population in US from 1955 to 2004.

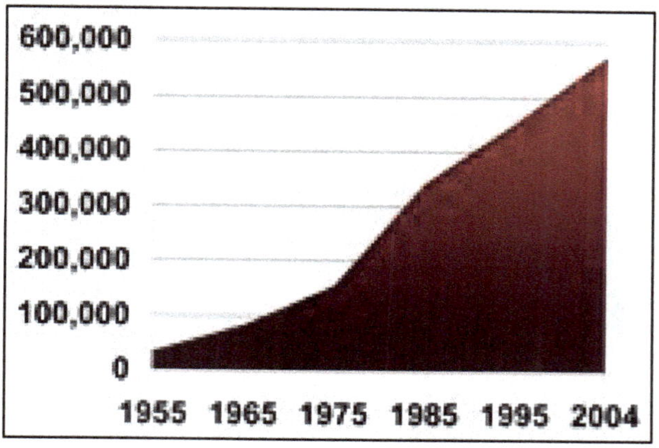

Figure 14.2: Foreign Students Population 1955 to 2004 (*Source*: Institute for International Education)

Edu-Tourism in Canada

More than 130,000 students go to study in Canada every year and among them more go to learn English or French. Foreign students bring a rich culture to the classrooms.

International Students Contribution to Canadian Economy

The Honourable Stockwell Day, Minister of International Trade and Minister for the Asia-Pacific Gateway, released (October 28, 2009) a study showing that international students contributed more than $6.5 billion to the Canadian economy in 2008. This is the first time a dollar figure has been attached to the economic impact of international students in Canada. "International students provide a significant boost to Canada's economy," said Minister Day. "Their presence helps create thousands of jobs and generates billions in revenue. Our government will continue its efforts to promote Canada as the destination of choice for international students by enhancing outreach efforts at our missions around the world." The findings are part of a report entitled *Economic Impact of International Education in Canada* prepared by Roslyn Kunin and Associates, Inc., for the Government of Canada. The report was commissioned by Foreign Affairs and International Trade Canada and released by Minister Day at a meeting of the Association of Universities and Colleges of Canada. The number of international students in Canada has more than doubled since 1998 to 178,000, and their presence provided employment for over 83,000 Canadians last year (Figure 14.3).

In November 2008, the Government of Canada, in collaboration with the provinces, announced the launch of a new branding campaign aimed at attracting more international students. It conveys the message that a Canadian education opens doors to opportunities. "In addition to attracting students to Canada, our government has also made tremendous strides in issuing study permits in key markets," said Minister Day. "We are seeing rising approval rates in places like China and India. We have also established policies and programs that help international students

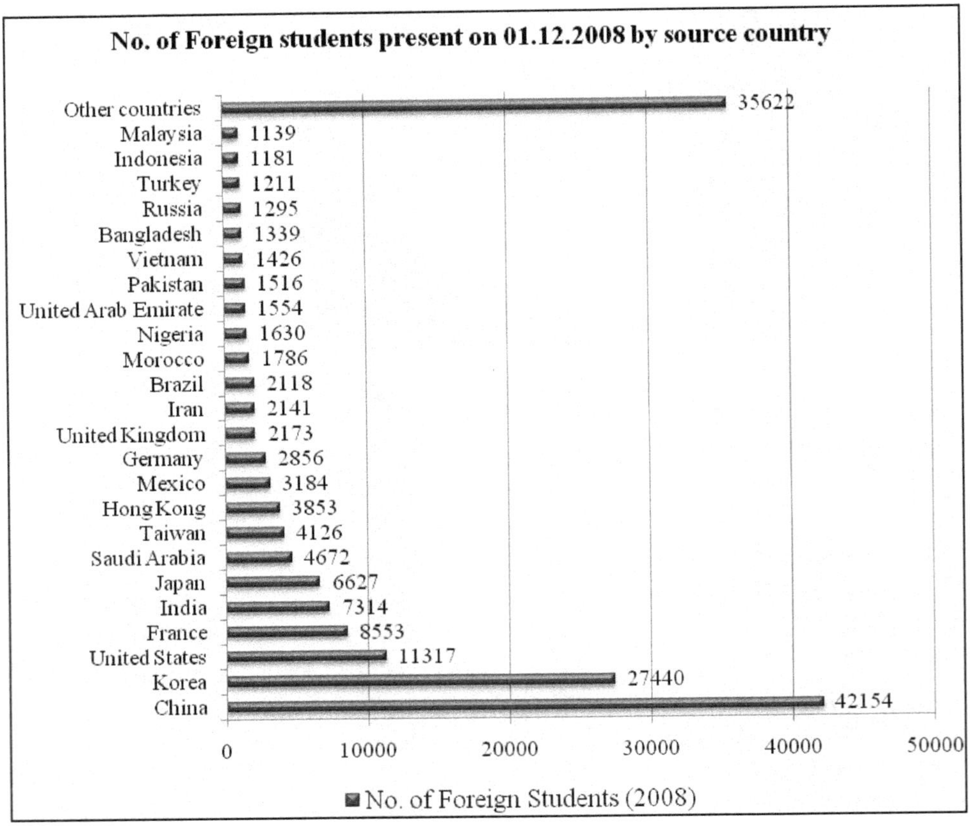

Figure 14.3: Foreign Students in Canada present on December 1st 2008 by Source Country (*Source*: www.cic.gc.ca)

gain valuable work experience during their study term and help meet skill and labor needs of Canadian employers." These efforts support the government's commitment to international education under its Global Commerce Strategy. Canada has seen a 7 per cent increase in international students since 2007. As shown in Table 14.3, China and Korea are the main sources of foreign students in Canada and in general, Asia is the most important source of foreign students in Canada.

Edu-Tourism in U.K.

Overseas students have become a lucrative source of earnings–contributing an estimated £10bn (almost $15bn) each year to the UK economy. The pursuit of overseas students has become big business–with UK universities competing with higher education institutions in countries such as the United States and Australia. International students provided a bigger source of income for UK universities in 2007/08 than government grants for research, the report adds.

Higher Education Minister, Bill Rammell said: "The significant increase in entrants from non-EU countries, up 10 per cent, clearly demonstrates that that the UK remains an extremely popular destination for international students. Recruiting fairs are held in Asian countries to promote interest in UK universities–and campuses of UK universities have opened in countries such as China and Malaysia.

Table 14.3: Foreign Students in Canada Present on December 1st 2008 by Source Country

Region	*No. of Foreign Students*	*Foreign Students per cent*
China	42154	24 per cent
Korea	27440	15 per cent
United States	11317	6 per cent
France	8553	5 per cent
India	7314	4 per cent
Japan	6627	4 per cent
Saudi Arabia	4672	3 per cent
Iran	2141	1 per cent
Other Countries	68009	38 per cent
Total	178227	100 per cent

Source: www.cic.gc.ca.

A report from the Organization for Economic Co-operation and Development (OECD) last autumn showed that the global market in overseas students had more than doubled in a decade. And there are forecasts from the British Council that the number of overseas students coming to the UK could triple.

The number of overseas students studying at UK universities has risen to 330,000–one in seven of the total. In 2001/02 there were only three institutions with more than 5,000 overseas students but by last year the rate had risen to 14. The university staff union, the UCU, warned that the growth in student numbers needed to be matched by increased investment in higher education. "Class sizes are extremely high and staffs are being forced to provide even more for less. The overlooked pastoral work of particular staff, often over and above their normal duties, is crucial to helping foreign students," said the union's general secretary, Sally Hunt.

The latest overseas total of 330,000 reflects a particular increase in students coming to the UK from India, Germany, France and Nigeria. This year's 15 per cent increase in students from India follows an increase of 14 per cent the previous year–with India now the second-biggest exporter of students to the UK. According to the study published recently, the number of students from outside the European Union has more than doubled in the past decade.

Despite a fall of almost 4 per cent, China remains the country with the single-largest number of students. Almost 51,000 Chinese students are taking courses at UK universities–accounting for almost one in six of overseas students (Table 14.4). One in 10 enrolments in 2007-08 – 229,640 students – was from outside the EU. In 1998-99, the figure was 117,290, which makes the increase 96 per cent. Students from outside the EU pay more for courses at UK universities than their European counterparts. In 2007-08, £1.88bn of UK universities' income came from non-EU students, while £1.76bn came from government research grants.

Table 10.4: Foreign Students in U.K. by Region of Origin, 2007

Region	*No. of Foreign Students*	*Foreign Students per cent*
China	51,000	15 per cent
India	47,000	14 per cent
Hong Kong	9,000	3 per cent
Other non-EU students	123,000	37 per cent
EU students(rather than UK)	100,000	30 per cent
Total	**330,000**	**100 per cent**

Between 2006-07 and 2007-08, universities saw a 10 per cent increase in income, the report says, partly due to the introduction of top-up fees and bigger grants. Professor Geoffrey Crossick, chair of Universities UK's longer-term strategy group, said: "This year's Patterns highlight the success of UK universities in challenging their counterparts in Europe, particularly in research and development and in attracting international students. Top 10 countries of origin are: China, India, Greece, Ireland, United States, Germany, France, Malaysia, Nigeria and Hong Kong.

Edu-Tourism in Australia

Australia has significant international people-to-people links, particularly with countries in the Asia–Pacific region. Each year, almost six million international travelers and students visit Australia for tourism or to undertake formal courses of study. They not only make a significant contribution to the Australian economy, but also develop important and often enduring links with Australia. Hundreds of thousands of young people who have studied in Australia and graduated from Australian educational institutions have become familiar with Australia and its way of life. Many continue to maintain personal and professional links with Australia long after they graduate in their chosen fields.

With around 5.6 million international tourists arriving in Australia in 2007, tourism has become Australia's biggest services export industry. In 2007 international visitors consumed around $23 billion worth of Australian goods and services. Tourism is an $81 billion industry for Australia and employs 464 500 people, 4.6 per cent of total employment. More than 90 per cent of tourism-related businesses employ about 20 staff, including substantial numbers of young people and people from non-English speaking backgrounds. Tourism is an important economic sector for regional Australia with 52 per cent of domestic tourism expenditure and 22 per cent of international visitor expenditure spent in regional areas. The international education industry is Australia's largest services export sector, contributing $11.7 billion to the Australian economy in 2006-07.

Australia is the preferred choice for international students from many countries and is the third most popular English-speaking study destination for these students.

In 2007, about 455,000 enrolments were recorded by students from more than 190 countries studying at educational institutions in Australia, including 5000 funded under Australian Government scholarships. Another 100,000 were studying Australian courses at off-shore campuses or by correspondence. Australia offers international students some 26,000 courses delivered by more than 1200 universities, training colleges and schools. The three biggest sectors in terms of enrolments were higher education (around 178,000), vocational education and training (121,000) and English-language intensive courses (102,000). Asia remains Australia's main source of international students, with more than 75 per cent, but enrolments from the Middle East, South America and Africa have grown strongly. International students are attracted by the high standard of Australia's education and training, its national qualifications framework, welcoming environment and diverse society. Australia's English-language schools provide a wide variety of training programs. These range from short courses to improve students' English-language skills while visiting Australia as part of a holiday, to formal courses in preparation for accredited levels of English recognized by education and immigration authorities around the world.

English-language training in Australia is provided by specialized institutions. A national accreditation body (the National ELICOS Accreditation Scheme) sets standards for class sizes, teacher qualifications, teaching methods and curriculum. Students from more than 100 countries come to Australia to learn english for academic, work, travel or personal reasons.

National legislation maintains high standards for those in Australia on student visas. Laws require registration of every course and institution that enrolls international students on student visas. International students can apply for permission to work part time during their stay in Australia. The Study in Australia website provides comprehensive information on options open to students, as well as a search facility covering all the courses and institutions available to international students. It also provides information about student visas. The website is available in numerous languages and a database lists scholarships by field and level of study. In 2007, according to Australian Education International(Table 14.5), the greatest number of international student enrolments in Australia came from the following countries: China; India; Republic of Korea; Thailand; Malaysia; Hong Kong; Japan; Indonesia; Brazil and the United States.

Edu-Tourism in Developing Countries

Edu-Tourism in India

With more than 17,000 colleges, 400 universities, 13 institutes of national importance and various other vocational institutes, the higher education system in India is one of the largest in the world. However, it is the fast integrating world economy and corresponding rise of students' mobility that have made studying in India an attractive option. There is large number of Indian as well as foreign students who apply every year to Indian universities and colleges. Higher Education in India is one of the most developed in the entire world. There has in fact been considerable improvement in the higher education scenario of India in both quantitative and qualitative terms. As a result, students from various parts of the world are coming today for higher education in India.

Table 14.5: Foreign Students in Australia by Region of Origin, 2007

Region %	*Countries*	*No. of Foreign Students*	*% of Foreign Students*
Asia (92%)	China	107,071	33 per cent
	India	63,604	20 per cent
	Republic of Korea	34,674	11 per cent
	Thailand	19,987	6 per cent
	Malasia	19,874	6 per cent
	Hong Kong	19,742	6 per cent
	Japan	16,077	5 per cent
	Indonesia	14,919	5 per cent
Latin America (4%)	Brazil	12,545	4 per cent
North America (4%)	United States	11,822	4 per cent
Total	320,315*	100 per cent	

Note: Percentages are rounded.

Source: Australian Education International.

* In 2007, about 455,000 enrolments were recorded by students from more than 190 countries studying at educational institutions in Australia, including 5000 funded under Australian Government scholarships. Another 100,000 were studying Australian courses at off-shore campuses or by correspondence.

Some of the advantages of studying in India are:

Low Cost

The cost of education in India is relatively low as compared to many other countries of the world.

Quality Education

Quality of education is not uniform throughout the length and breadth of India. However, there are some educational institutes in India that provide world class education. Indian Institutes of Technology, Indian Institutes of Management, Indian Institutes of Science, National institutions and central universities like Pusa, NDRI, Jawaharlal Nehru University, Aligarh Muslim University, Banares Hindu University and Delhi University are some such Institutes. The government of India is also speeding up the efforts to establish more such institutes that can offer quality education in India.

Financial Assistance

Various scholarships, education loans and other financial assistance are now available for studying in India today.

Consultation Service

The government of India provides consultation service to the interested international students through Education Consultants of India (Ed.CIL). Thus one can get all the information about the Indian education system, cost of education, duration, visa, accommodation facilities even before landing up in India.

Unique Courses

Apart from above mentioned advantages, one can also study some unique courses that were discovered and developed by the traditional knowledge system of India. Ayurveda, Sanskrit, Yoga, Hindi, Petroleum and Energy are some such courses that enthuse many international students.

As the competition from foreign universities come calling in the home turf, Indian universities have joined the rat race of wooing the foreign students. Now, Indian universities have witnessed an increase in the fresh enrolment of international students in domestic varsities from 13,267 in 2004-05 to 14,456 in 2005-06. The study also points out about 80 percent of international students were enrolled in undergraduate programmes. Delhi University Vice-Chancellor Deepak Pental said most international students want to pursue economics and commerce.

Data collected by the AIU from 1992-93 to 2003-04 suggests that the number of international students coming to India has steadily increased during the first half of the 1990s, with a peak of over 13,000 achieved in 1993-94. "This trend was probably due to the fact that while many developed countries, and especially the United Kingdom and Australia, were aggressively marketing their educational ware, India was inactive in this regard," states a research paper by Veena Bhalla, an AIU member. In 2005-06, most students came from West Asia, followed by South and Central Asian countries. The UAE sent the maximum students to India for higher education. In 2004-05, the UAE sent 1,500 students, which rose to 2,034 in 2005-06, Neighboring Nepal followed closely with 1,411 students, while 1,264 students from Iran came to India to pursue a degree. Adding to the quality of education is the fact that English is the generally accepted mode of teaching and living standards are economical–attracting thousands of students from at least 100 countries.

A large number of students come to India from countries like Afghanistan, Bangladesh, Bhutan, Ethiopia, Fiji, Iran, Iraq, Japan, Jordan, Kenya, Ghana, Lebanon, Madagascar, Malaysia, Mauritius, Myanmar, Nepal, Somalia, Sri Lanka, Suriname, Syria and even the US.

Edu-Tourism in Malaysia

Malaysia is located in the tropics of Southeast Asia, straddling the South China Sea. Recent developments are putting Malaysia on the world map, as she focuses on becoming a knowledge-based economy. To develop world-class quality education and meet the demands and requirements of the new millennium, as well as to affirm the position of English as a second language, the Government has initiated major educational reforms by formulating new legislations on education. These were tabled in the Parliament from 1995 to1997, and cover all levels of education, from pre-school to higher tertiary education. Malaysia is indeed becoming a dynamic and exciting regional centre for education excellence.

Malaysia's leading newspaper 'The Star' reported that figures had increased between 2006 and 2008 by 30 per cent, bringing the overall numbers of international students in Malaysian international schools and higher education institutions to 65,000. According to the following calculations by industry analyst (see pamjitsingh.ppt) the Malaysian government is well on target to realize its 2010 goal of 100,000 international students. Taking into account the forecast in world demand by 2010, the Malaysian government estimates that their market share would need to grow from its current world share of international students (schools and higher education) of 3.9 per cent in 2004 to 6.6 per cent in 2010. In comparison to the global average annual growth rate of international students which is around 7.4 per cent P.A, the Malaysian target growth rate would need to be in the order of 24.0 per cent per annum to achieve the 2010 target. In order to realize this goal, a new Higher Education Ministry Marketing and International Education Division was created.

Branding has emerged as an important strategy for governments seeking to strategically develop their higher education markets. The 'Malaysian Education' brand draws on deep cultural, religious and political resonances to promote its product – one that emphasizes lifestyle, culture and quality of education. This includes the value to be gained from its unique multicultural population of Malay, Indian and Chinese; its Islamic religion; and its experience of colonialism. Despite the contradictions inherent in this new form of neo-colonialism, these cultural values and symbols are being effectively mobilized to open up the African, Arab, Chinese and Indonesian markets. Malaysia's story demonstrates the high level of fluidity in globalizing the higher education market. It requires players to be highly competitive, constantly utilize intelligence, be attentive to strategies as to how to open new markets, and have a way of representing the sector as an attractive and unique brand. The Higher Education Ministry of Malaysia revealed that expecting about 72,000 foreign students selected Malaysia as their destination in 2009 and they believe that more foreign students may choose to pursue their tertiary education in Malaysia because of the world recession and the higher education ministry of Malaysia is pushing to make the country the hub for higher education excellence in the region; The figure was a 40 per cent increase compared with 2007. Most of the students are from China, followed by Indonesia and the Middle East.

Results and Discussion

It is clear that Asia is the Key source region for educational tourism in developed and developing countries; Asian countries cannot meet domestic demand for higher education, and a rising middle and higher class in those nations due to rising economy means more families have the money to send their children overseas to study and they look for the best option according to their budget and objectives, whether to send their children to US., Canada, Australia, UK or any other developing country such as India, Malaysia and etc. On the other side, there are student mobility between developing countries, for example, 4 per cent of foreign student of Australia are from US or 6 per cent of foreign students of Canada are from US, or 30 per cent of foreign students in UK are from other European countries and such mobility does not only rely on the need for higher education but it also could be interpreted that students do not only need higher education, but they want to experience the culture of other countries, they want to live in a new situation independently and many other reasons which makes them to look beyond education.

According to the Table 14.6, out of 1,579,387 foreign students studying in US, Canada, UK, Australia, India and Malaysia; the market share of US is 43 per cent followed by UK 21 per cent, Australia 20 per cent and Canada 11 per cent (the average market share of four developed countries is 24 per cent where as the average market share of two developing countries is 2.5 per cent)and the market share of India and Malaysia as Asian developing countries is only 5 per cent but on the other side, Asian countries are the main source of foreign students in developed countries! What could be the reason behind it? It can be interpreted that education system of developed countries is highly developed and the main reason is their policy on attracting foreign students which highly relies on marketing of their education system specially their higher education; developed countries has realized the importance of edu-tourism and revenue generation of this sector, they could realize the booming student mobility and demand for higher education and they do emphasis on attracting consumers(students) from Asian countries due to the high population of Asia and increasing income level of Asians and willingness for investing on the education and future of their children.

Table 14.6: No. of Foreign Students in each Country and their Contribution to the Economy

	Country	*No. of Foreign Students*	*Market Share*	*Avg. Market Share*	*Contribution of Edu-Tourism to the Economy (billion$)*
Developed	United States	671,616 (2008-2009)	43%	24%	17.6
Countries	United Kingdom	330,000 (2007)	21%		15
	Australia	320,315 (2007)	20%		11.7
	Canada	178,000 (2008)	11%		6.5
Developing	Malaysia	65,000 (2008)	4%	2.5%	<1
Countries	India	14,456 (2004-2005)	1%		<1
	Total	**1,579,387**	**100%**		

Conclusion

Based on above discussion, due to globalization and the rise of economy and the increase of spending power of people in developed as well as developing countries, student mobility for educational purpose has created a huge market for edu-tourism and revenue generation which should be considered by countries like Iran that has risen to the rank of 22 in scientific production in 2009 (which can be a great resource for edu-tourism) and is one of the first 10 top countries in tourist attractions (great destination for eco-tourists) and has more than 1,200,000 historical attractions and one of the most important religious hubs. Therefore with all the potential, attractiveness and advantages, Iran can attract many International students from border neighbor countries, Middle East, Africa, India, China and other countries and the revenue generation of this industry can exceed what Iran earns today by selling oil which is a limited resource; but to achieve this goal, there must be some strategies to improve the infrastructure and policies to facilitate the attraction of International students. By looking at the figure of edu-tourism in developed and developing countries and the increasing market, it is obvious that this sector highly contributes in sustainable economic development and determines an opportune time for Iran to develop new approaches that combine unique aspects of the local environment with learning/ education such as cultural activities (*i.e.* language, religion, art, music, or history studies), environmental or eco tourism activities, medical tourism, education programs in hotels, conference, seminars, incentive programs and formal university programs. Some of the marketing strategies which Iran can take to improve edu-tourism industry are as followings:

- Opening of new edu-tourism desk in all Iranian embassies and consulates and utilizing them as marketing agents for edu-tourism facilities in Iran.
- Organizing exhibitions at important source countries by participations of various Iranian Universities and educational institutions to show their potential of high quality education and attract foreign students;
- Establish an internet portal in various languages for global consumers;
- Designing an aggressive and powerful marketing strategy and to create an effective marketing plan of actions, availability of edu-tourism information through various media;
- Publication of brochures, posters, picture postcards and CDs and their distribution through various science and education departments of source countries.
- Production of short films on Iranian universities and showing their national and international achievements and also the high quality of life style if Iranian in metro cities by screening these films in various educational and cultural programs of source countries.

References

Bodger, D., 1998. Leisure, learning, and travel. *Journal of Physical Education*, 69(4): 28–31.

Brent W. Ritchie, Neil Carr and Christopher P. Cooper, 2003. *Managing Educational Tourism*, p. 283.

Conrady, R., 2010. *Trends and Issues in Global Tourism*, p. 134–175.

Harris, R., Tony Griffin and Peter Williams, 2002. *Sustainable Tourism: A Global Perspective*, Butterworth-Heinemann, p. 214–256.

http://aei.gov.au/AEI/PublicationsAndResearch/Snapshots/Default.htm (accessed 05 Nov. 2010)

http://aei.gov.au/AEI/Statistics/Default.htm (accessed 03 Nov. 2010) http://mospi.nic.in/Mospi_New/site/India_Statistics.aspx?status=1&menu_id=14 (accessed 02 Nov. 2010)

http://news.bbc.co.uk/2/hi/uk_news/education/7181806.stm (accessed 02 Nov. 2010)

http://news.bbc.co.uk/2/hi/uk_news/education/8588779.stm (accessed 03 Nov. 2010)

http://stats.oecd.org/Index.aspx (accessed 02 Nov. 2010)

http://unwto.org/facts/menu.html (accessed 02 Nov. 2010)

http://ur.isc.gov.ir/Shownews.aspx?n_code=23 (accessed 02 Nov. 2010)

http://www.asiaone.com/News/Education/Story/A1Story20090405-133398.html (accessed 03 Nov. 2010)

http://www.cic.gc.ca/english/resources/statistics/facts2008/temporary/09.asp (accessed 02 Nov. 2010)

http://www.guardian.co.uk/education/2009/sep/24/internationalstudents-students (accessed 05 Nov. 2010)

http://www.infoplease.com/ipa/A0762380.html (accessed 03 Nov. 2010)

http://www.international.gc.ca/media_commerce/comm/news-communiques/2009/319.aspx (accessed 05 Nov. 2010)

http://www.photius.com/rankings/population/birth_rate_2010_0.html (accessed 01 Nov. 2010)

http://www.photius.com/rankings/population/education_expenditures_percent_of_gdp_2010_0.html (accessed 02 Nov. 2010)

http://www.photius.com/rankings/population/population_growth_rate_2010_0.html (accessed 02 Nov. 2010)

http://www.realonlinedegrees.com/education-rankings-by-country/(accessed 05 Nov. 2010)

http://www.telegraph.co.uk/news/uknews/1563009/Number-of-foreign-students-in-UK-rising.html (accessed 02 Nov. 2010)

http://www.timeshighereducation.co.uk/world-university-rankings/2010-2011/top-200.html#score_OS | sort_rank | reverse_false (accessed 03 Nov. 2010)

http://www.unfpa.org/public/(accessed 02 Nov. 2010)

Kandari, O.P. and Chandra, Ashish, 2004. *Tourism, Biodiversity and Sustainable Development*, p. 4–14.

Kulkarni, M.V., 2005. *Tourism Marketing*, p. 55–57.

Misra, S.N. and Sadul, S.K., 2008. *Basics of Tourism Management.*

Prasad, V. and Sundari, V., 2009. *Travel and Tourism Management*, p. 241–242.

Smith, C., 1982. *Education Tourism: Travel and Tourism Analyst*, 3: 60- 75.

Chapter 15

Development and Optimal Utilization of Nigeria's Natural Resources for Industrial Growth: Case Study of RMRDC's Catalytic Model Plants

A.M. Wele

Raw Materials Research and Development Council,
Nigeria
E mail: wele92@hotmail.com

ABSTRACT

The paper highlights a programme of Raw Materials Research and Development Council (RMRDC), Nigeria, aimed at establishment of catalytic model factories based on local availability of raw materials, for the purpose of stimulating the emergence of small and medium scale enterprises. The catalytic plants were used to enhance indigenous capacity for equipment and machinery design, fabrication and encouragement of the private sector towards utilization of proven indigenous engineering works for raw materials processing. It also stimulates the sector towards the establishment of local resource -based industries, taking advantage of the wide range of incentives that has been put in place by the Nigerian government.

Keywords: *Raw Materials Research and Development Council, Catalytic model plants, Local resource based industries, Natural resources, Indigenous engineering capacity.*

Introduction

Nigeria is blessed with numerous natural resources. These can be classified into agricultural, mineral, water and marine, renewable energy, and human resources.

Each of these has the potential of contributing more than oil and gas towards revenue generation in particular and national development in general.

However, the country still imports up to 90 per cent of its industrial raw materials required for production. This, among other factors makes Nigerian products uncompetitive. This situation, in turn, encourages imports of almost every product into Nigeria. This uncontrolled importation has also worsened the situation of the industries, which has lead to low capacity utilization, outright closures and relocation of many industries to other countries with more favorable economic environments. The social consequence of this is unemployment and poverty.

In order to reverse this trend, there must be a judicious application of Science, Engineering, Technology and Innovation (SETI) to the production or manufacturing process, especially in converting natural resources to industrial raw materials. Of the four important factors above, engineering skill is the most important. This is because engineering provides the design and fabrication skills required to develop processes and process equipment required for value addition. It also provides the necessary engineering infrastructure required for a robust manufacturing sector including machine tools, machine components, industrial plants, power systems, transport systems and the intellectual skill for installation, operation and maintenance of machines.

It is recommended that for Nigeria to achieve its desired goals in the year 2020, it must increase the local content in industrial raw materials from the present 5 – 15 per cent to at least 70 per cent, by the year 2020. Efforts towards this can be concentrated on resource based industries where Nigeria has comparative advantages. The strategies for achieving this include: establishing a robust National System of Innovation (NSI), implementing the new National Raw Material Policy, establishment of a National fund for innovation and competitiveness for industries, establishment of a University of Natural Resources, establishment of Raw Materials Processing Cluster (774 – *i.e.* one in every local government), establishment of engineering skill and industrial parks next to faculties of engineering in higher institutions and adequate funding for research in engineering applications in Raw Materials Processing. It is my belief that if these and other interventions are implemented, it can lead to the emergence of thousands of engineering based industries producing most of the required raw materials for local industries and indeed competitive finished products. This can make Nigeria a global player in manufacturing and most of the challenges confronting the country today can be overcome.

Government has over the years attempted to address the challenges. This has led to the establishment of a network of institutions including institutions of higher learning with engineering programmes, research institutes and development centres. In addition, the private sector has also organized themselves into groups such as Manufacturers Associations of Nigeria (MAN,) Nigeria Association of Small Scale Industrialists (NASSI,) Nigeria Association of Small and Medium Enterprises (NASME,) and Chambers of Commerce. Industrial policies have also been established. These interventions have yielded results with varying degrees of success. For example, the Raw Materials Research and Development Council (RMRDC) has developed many technologies for value addition to natural resources, among which are the catalytic model factories. The Council has also promoted investments and skill acquisition in raw materials development.

The Raw Materials Research and Development Council (RMRDC) Nigeria, a para-statal organization under the supervision of the Federal Ministry of Science and Technology, was established following the need to add value to local raw materials as well as develop them into industrial inputs for the manufacturing sector, with the ultimate objective of catalyzing industrialization, generating employment, creating wealth and reducing poverty.

The primary mandates of the Council are:

- ☆ To draw up policy and action programmes on raw materials acquisition, exploitation and development;
- ☆ To review from time to time, raw materials resources availability and utilization with a view to advising the Federal Government on the strategic implication of depletion, conservation or stock-piling of such resources;
- ☆ To advise on adaptation of machinery and process for raw materials utilization;
- ☆ To provide special research grants for specific objectives and device awards or systems for industries that achieve breakthrough or make innovations and inventions; and
- ☆ To encourage the publicity of research findings and other information relevant to local sourcing of raw materials.

The Council executes the following core programmes to provide a strategic framework for the actualization of its mandates:

- ☆ Funding of Research and Development (R&D) on the development of raw materials as well as machinery and equipment for value addition.
- ☆ Boosting of the production of industrial crops in order to enhance availability of raw materials for the industrial sector.
- ☆ Capacity Building and awareness creation on raw materials availability, as well as availability, development and adoption of modern technology, R&D output etc. for value addition on local raw materials.
- ☆ Bilateral and multilateral programmes to foster partnerships, collaborative research and technology development and promotion programmes on raw materials development.
- ☆ Investment promotion in resource-based industries for production of raw materials.

Value Addition to Raw Materials

- ☆ Value addition to agricultural produce and solid mineral resources through promotion of processing ventures and process technologies.

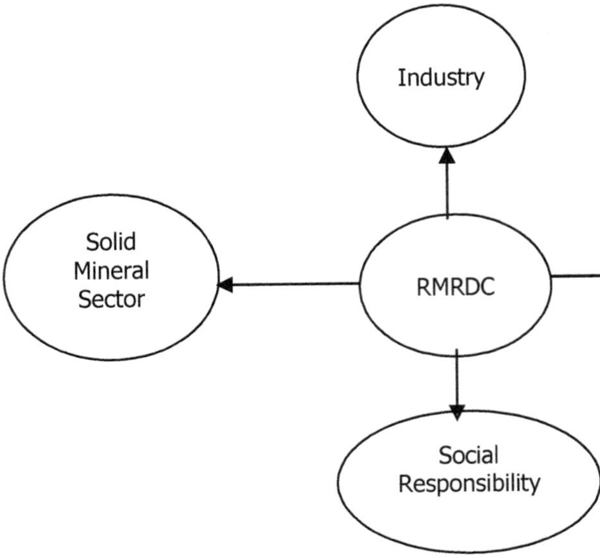

Figure 15.1

Agricultural Raw Materials

- The Council is linked to agriculture in terms of promoting processing of agro commodities in order to generate raw materials in terms of quantity and quality that will serve as feedback for industries.
- Carry out studies to identify available processing technologies and technology gaps and address such gaps.
- Generate information from the studies and promote investments in agro processing.
- Conduct research studies on availability of agricultural raw materials for industrial utilization.
- Identify available process technology both indigenous and foreign, as well as identify gaps.

Expected Economic Impact

- Increased non-oil export.
- Increased level of production by rural farmers thus increasing income and addressing poverty.
- Establish a reliable raw materials production base to attract investment in Agro commodity industries in order to produce value-added products for the export and local utilization.
- Create wealth and employment by stimulating the establishment of downstream processing industries from these commodities.

- Promotion of R&D on the development of exotic breeds to facilitate increased milk yield from local cows to attract the establishment of dairy industries.
- Upgrading of indigenous technologies to enhance increased domestic utilization of raw materials to produce added value products for export.
- Support a national study on medicinal herbs in order to identify raw materials for the pharmaceutical industry.

Mineral Raw Materials

- The Council is linked to this sector through the promotion of processing of solid mineral resources currently in their natural form.
- Carry out studies to generate information in collaboration with relevant agencies in the sector.
- Identify existing technologies and technological gaps in the beneficiation of solid minerals and general information to promote investments in the sector.
- The Council intervened in the sector to promote value addition to solid minerals in order to produce intermediate raw materials to serve as feedstock for industries.
- Collaborates with Geological Survey, Research Institutes, Universities and the Organized Private Sector to conduct studies on solid mineral raw materials availability.
- Promote the development of some key solid mineral raw material by catalyzing the establishment of resource-based industries:
- Provide technology innovation to enhance the output of small scale miners in the country.

The Techno-Economic Survey of Resources and Technologies has remained one of the programmes of the RMRDC since 1988/89, when it was first conducted. Subsequent updates carried out until 2009 ensured that the 10 industrial sectors of the Manufacturing Association of Nigeria (MAN) were covered. The programme aims at gathering data on raw materials available in the country, their locations and reserves as well as their various levels of exploitation, development and utilization.

Agricultural Commodity Surveys on Cassava, Cotton, Cocoa, Groundnut, Gum Arabic, Maize, Rubber, Palm produce, Soyabean, Cashew, Mango, Ginger, Castor, Pineapple, Citrus fruits, Sugarcane and Sorghum. The reports of these studies are made available to the public and stakeholders through publications, Council's website and other electronic means.

Surveys of Solid Minerals were also carried out as contribution to the efforts of other organizations and strengthen Government's focus on this sector. Data obtained from these surveys were presented in pamphlets as *Technical Briefs.*

Inventory of Machinery and Equipment Fabricators conducted across the country aimed at identifying designers and fabricators of machinery and equipment. The data obtained was presented in a publication made available to the general public to facilitate patronage and collaboration. These various surveys became the source of providing up-to-date information to industrialists, investors, researchers, policy makers etc.

The Council commenced the publishing of a periodical, titled: *Journal of Raw Materials Research*. Four editions of the journal have been published. It has offered a platform for researchers to publish works on local raw materials in Nigeria.

Raw Materials Research and Development Council's (RMRDC) Catalytic Model Plants

Findings in the early years of operation of the Council showed a slow response of the private sector to the call for the establishment of local resource-based industries despite government incentives. The Council designed a programme to stimulate, create and elicit interest of prospective investors towards resource-based investments, and also provides industrial extension services to clients to facilitate the smooth translation of project ideas into commercially profitable industrial ventures. Indeed the promotion of resource-based industries is seen as a strategy towards the implementation of the National Economic Empowerment Development Strategy of the Federal and State Governments in Nigeria.

To overcome this constraint, RMRDC initiated catalytic model projects and joint venture programmes with the objectives of demonstrating to stakeholders that the technologies for processing various raw materials are available locally and to create awareness on the profitability of such ventures. Other objectives of the programme includes the encouragement of the private sector to utilize proven indigenous engineering works for raw materials processing and to stimulate the establishment of local resources- based industries, taking advantage of the wide range of incentives that were put in place by the Nigerian government. The programme was also aimed at raising the supply of locally processed raw materials to Nigerian industries and the eventual savings in foreign exchange.

The projects so far promoted include the following:

1. Talc Processing Plant.
2. Soda Ash Production Plant,
3. Phosphate Processing Plant.
4. Pharmaceutical/Industrial grade Kaolin Projects
5. Cocoa Processing Plant.
6. Granulated Limestone Processing Plant.
7. Hydrated Lime Production Plant.

Talc processing plant

Talc is a natural and rich mineral in the world with various colors of white to pink, light green, grey and black. Talc is also named as soapstone, French chalk and steatite. It is a hydrated magnesium silicate deposited by the metamorphosis of dolomite and quartzose rocks or by the hydrothermal alteration of ultramafic and mafic rocks.

The chemical molecular formula is $Mg_3Si_4O_{10}(OH)_2$ with hardness of 1 to 1.5, and a specific gravity of 2.7 to 2.8. Talc is the softest mineral with extreme softness, smoothness, slippery and lubricating, chemical inertness, high luster and sheen, high fusion point, low heat and electricity conductivity and pigment extenders.

Talc is an extremely useful and economically important mineral. It is widely used in ceramics, paint, paper and plastics, agriculture and food, personal care, mold release agents, wastewater treatment, textiles, etc.

The five main characteristics of talc that make it a unique material for industrial and domestic applications are:

1. Lamellarity (made of platelets that easily slide on each other)
2. Softness (unctuous and not abrasive)
3. Chemical inertness
4. Affinity for organic chemicals
5. Whiteness

An estimated reserve of over 100 million tons of talc has been known to exist in Niger, Osun, Kogi, Kwara, Ogun, Taraba, and Kaduna states. There are only two medium processing plants currently operating in Nigeria, and both are in Niger state. The color of the Nigerian talc varies from white through milky -white to gray. The talc industry represents one of the most versatile sectors of the industrial minerals of the world. The exploitation of the vast talc deposit in Nigeria would therefore satisfy not only local demands but also that of the international market as well.

This project is very significant to the national economy as talc has wide industrial application in the pharmaceutical, cosmetic, rubber and ceramic industries. The average annual national demand for talc is conservatively put at 50,000 metric tons. Currently, pharmaceutical grade talc is not being produced locally, even though about 40 million metric tones of proven raw talc is available in the country.

The current demand of some of these industries in Nigeria is shown in Table 15.1.

Table 15.1: Talc Consumption by some Industries

Industry	*Talc requirement (Tones/year)*
Paint	10,000
Paper	10,000
Ceramic	10,000
Cosmetic	5,000
Soap	2,500
Rubber	2,500
Insecticide and fertilizer	10,000

The Raw Material Research and Development Council's model factory is located at Kagara, Niger state Nigeria, with an installed capacityof 3,000 metric tons/annum.

All equipment installed for the project was fabricated by local equipment fabricators. The equipment fabricated was: jaw crushers, hammer mills, pebble mills, classifiers, slurry pulps etc. The plant commenced production in December 1994.

Soda Ash Production Plant

The occurrence of natural soda reserves, reported from several places in the country can be classified into two categories, *i.e.* those occurring in the form of Sodium mineral salts, such as TRONA ($Na_2CO_3NaHCO_32H_2O$), and those occurring in the form of BRINE, a solution containing Carbonate, Sulphate, Sulphite and Sodium salts.

Soda Ash is produced in the form of white hygroscopic powder or as granule, and it can be used as a final product in that form. Soda Ash with bulk density of less than 0.8 per cent per cent is called light soda ash. Recent trends indicate increase in the demand for the Soda Ash of higher density of approximately 1.2 or more, and also having larger particle sizes.

Soda ash or sodium carbonate is an inorganic chemical with a number of industrial applications. The glass industry represents the major end-use segment for soda ash. Soda ash is also used in the manufacture of chemicals, as a builder in the manufacture of detergents, in the pulp and paper industry, and on a smaller scale, for some other uses such as water treatment.

Soda Ash: Raw Material (Trona) Deposits in Nigeria:

Trona, a sodium solid mineral which is the main solid raw material for the production of Soda Ash (Sodium Carbonate) is hugely deposited in the Northern States of Nigeria. It is found in Borno, Taraba and Adamawa States. It is also extensively deposited in Manga, Tula, Geidam, Yusufari, Kukawa in Yobe state, Kazaure in Jigawa state and Lake Chad basin.

The brine solution which contains carbonate, sulphate, sulphite, and sodium, used also as main raw material in solution for the production of Soda ash (sodium carbonate), is also found in Nigeria in the under listed states.

- Benue State in Gboko, Guma, Gwer, Ado and Katsina-ala LGA.
- Ebonyi State in Ikwo LGA
- Taraba State in Karim-Lamido LGA

Future consumption of caustic soda, soda ash, and by-products, by the water industries, environmentalist, pulp, oil and gas drilling, oil refining, glass industries, food industries, paper, chemical industries, and other end users, is highly likely to geometrically increase due to industrial growth, as such, the international production tonnage in developed countries have been predicted will as well geometrically increase as to meet up with the on coming drastic demand.

This project which is a joint venture between Raw Materials Research and Development Council and a State in Nigeria has an installed capacity of 3,000 metric tons of soda ash (sodium carbonate) per annum. It was commissioned on 26th July, 1995. The equipment fabricated locally were: crushers, pulverisers, rotary calciners, dissolvers, sedimentation tanks, filter press, evaporators, crystallizers, Steam boilers, conveyors, etc.

Phosphate Processing Plant

Phosphate rock occurs in pebbles in a matrix of mud and silica sand. Phosphate rock deposits occur in large quantities in Sokoto and Ogun states in Nigeria. There is also occurrence of the rock in parts of Imo and Edo states. Phosphate is an essential raw material for the production of phosphate fertilizers. It is also used in the production of phosphoric acid.

Although the discovery of rock phosphate in Nigeria dates back to 1921, its exploitation only began to receive attention in 1990 when RMRDC and a state in Nigeria embarked on a phosphate processing plant.

The project has two production lines producing 15,000 tones of refined phosphate rock per annum and is located in Sokoto state, Nigeria. The contract for the designing, fabrication and installation of the plant/machinery was awarded in May, 1991. While by 1st quarter 1993, all infrastructures were in place and the production line had been installed and ready for test running. The production line which is based on the wet milling method was found during trial production, to be consuming too much energy and some of the units of equipment were deficient, which necessitated some modifications in 1997.

The significance of this project to the Nigerian economy is due to the fact that, the national demand for phosphate is put at 200,000 metric tons, all of which is imported by the chemical and fertilizer industries in Nigeria. This prompted the need for the country to look inward with regards to sourcing of major industrial raw materials locally. The equipment locally fabricated and installed were: rotary washers, jaw crushers, vibrating screen, hydraulic classifiers, floatation machine, silos, hammer mills with classifier, bagging equipment etc.

Pharmacuetical/Industrial Grade Kaolin Projects

Kaolin is a versatile industrial mineral of wide application in the cosmetic, pulp and paper, chalk, agro-allied and pharmaceutical industry. The cosmetic and pharmaceutical industries in Nigeria have shown general apathy towards local sourcing of kaolin. Apparently, this low interest stems from the stringent quality of kaolin required by these industries. More so, the locally sourced kaolin is speculated to be heavily laden with microbes. As a result, huge quantity of kaolin is continually being imported into the country annually to meet the pulp and pharmaceutical need.

The total consumption of pharmaceutical grade kaolin in Nigeria is about 20,000 tons/annum. Medically, kaolin is used as an absorbent and has also found tremendous use in anti-diarrhoea formulations. It is sometime used in combination with pectin, a polygalactunic acid extracted from the pulp residues of citrus fruits. Conservatively about 500 metric tons of kaolin are required annually for the formulation of anti-diarrhea alone. Thus local sourcing of kaolin would not only enhance the input of the pharmaceutical industries, but also conserve some foreign reserve for Nigeria.

Kaolin deposits are available in large quantities in states of Jigawa, Kano, Kebbi, Sokoto, Ogun, Imo, Kogi, Rivers Anambra Zamfara etc.

The Raw Materials Research and Development Council promoted the development of two Kaolin processing plants in Nigeria. The first project was the Pharmaceutical grade Kaolin plant located at Kankara in Katsina state, Nigeria in 1997, while the second was the Industrial grade kaolin plant located at Gwarzo, Kano state in 1998. Both plants were designed to use wet processing methods for the production of pharmaceutical and industrial grade Kaolin respectively. The equipment fabricated locally were: blungers, hydrocyclones, sieves, tanks, filter presses, dryers, hammer mills, floatation machines, weighing and bagging machines.

Cocoa Processing Plant

Cocoa is an important raw materials required in the production of semi-finished products such as cocoa powder and butter. Cocoa powder is obtained from cocoa cake. The cake is formed as a result of processing the butter from processed cocoa beans. Mechanically, the cakes are crushed into small pieces called kibbled cake through pulverizes and air separators to produce a fine powder. Cocoa powder is mainly used for the manufacture of beverages. Powder with residual cocoa butter is used in the production of cookies, biscuits, chocolate, syrups and spreads and vegetable fat coatings.

Cocoa butter is obtained from cocoa powder and it's used in the production of chocolates and body creams. Cocoa butter is a stable natural fat which easily melts just below 33 degree Celsius. An estimated annual demand for cocoa powder/butter in Nigeria is about 250,000 metric tons/annum. Currently only about 100,000 tons are produced in the country. With a short fall of 150,000 metric tons, about 14 small and medium scale plants of 11,000 tons/annum are required to be established to meet the national demand.

The main raw material which is Cocoa beans are grown extensively in the southern states of Nigeria namely: Ogun, Oyo, Osun, Ondo, Ekiti, Akwa Ibom, Cross rivers states

The Cocoa processing plant was established at Idanre, Ondo state, Nigeria. The equipment, like in the other projects, was fabricated locally.

Granulated Limestone Processing Plant

Limestone is widespread in the sedimentary basins of Nigeria. Prominent reserves have been proved to exist all over the country. There are over 85 million tones in Benue state, 70 million in Kwara, 100 million in Sokoto, 1 billion tones in Anambra, 100 million in Imo, over 7 billion tones in Ogun states. Deposits of commercial value have been discovered in Edo, Bauchi, Niger states and the Federal Capital Territory of Nigeria.

Generally, the supply of granulated limestone to end users has not measured up to the demand in Nigeria. Only about 30,000 tons out of the estimated 300,000 tons per annum of the national requirement is met by the combined out put of local producers.

Apart from its uses in the fertilizer industry, granulated lime stone has found enormous uses in the glass, paper, paint, ceramic, refractory chemicals, plastic rubber and pharmaceutical industries. There is, therefore, a good investment prospect for this non-metallic mineral.

As part of its contribution to the national efforts to increase agricultural output, RMRDC consider that the use of composite fertilizer containing appropriate ration of lime or limestone could slow the rate of soil infertility and possibly reverse the current trend with time. The establishment of a plant to manufacture granulated limestone for use by fertilizer blending companies is thus expected to have a multiplier effect on the nation's agriculture.

The basic raw material for the production of granulated limestone is good quality limestone. Benue, Kogi, Sokoto, Anambara, Imo, Kogi, Ogun, Edo, Bauchi, Cross River States have millions of tons of good quality limestone.

The entire equipment required for the granulated limestone production was fabricated locally. The plant has an installed capacity for the production of 20,000 tons of granulated limestone per annum.

The production process employed in the Calabar plant involves very simple operations requiring the following components:

- Lime stone ore bin
- Jaw crushers
- Hammer mill
- Conveyer belt
- Classifier
- Sieve
- Granulator
- Bagging Machine

The Council in conjunction with a private entrepreneur went in to a joint venture arrangement and incorporated a company located in Calabar Cross River state for the processing of granulated limestone for fertilizer blending purposes. The plant has an installed capacity of 20,000 tons of granulated limestone/annum.

Hydrated Lime Production Plant

Hydrated lime is one of the most extensively used industrial chemicals. Its application in the chemical industry includes the production of caustic soda, precipitated calcium carbonate, hypochlorite bleaches and fertilizers. In the food industry, it is used for water softening and sugar refining. Lime is also used for agricultural and metallurgical purposes. Current national demand for lime is estimated at about 500,000 tons while supply is about 135,000 tons per annum. The raw materials required for lime production (limestone) is available in Cross River, Delta, Ogun, Enugu, Niger, Benue and Kogi states etc.

The RMRDC project located at Kwakuti, Niger state has an installed capacity of 6,000 tons of hydrated lime/annum.

Conclusions

Industrial development is a major parameter for measuring the development of advancement of a country's economy. The strength and health of the manufacturing industry is largely dependant on the availability of raw materials and the efficiency of research and development activities which fall within the mandates of RMRDC. The Council vision of becoming an indispensable catalyst for industrial growth and development in Nigeria is clearly in consonance with the Federal Government's goal of making Nigeria one of the twenty largest economies in the world by the year 2020.

The imperative of this coincidence of vision is that the Raw Materials Research and Development Council have a serious role to play in the development and growth of the Nigerian economy.

The heavy dependence of the economy from the export of petroleum and gas products makes the Nigerian economy highly vulnerable. It is therefore obvious that, the future of Nigeria depends strongly on the diversification of the economy with emphasis on the expansion of the agricultural, mineral and consequently the industrial manufacturing sectors, pivoted on increased utilization of the nation's raw materials. This again is in tune with RMRDC's mission, which is to promote the development and optimal utilization of Nigeria's natural resources for industrial growth. The catalytic model plants established by the council have impacted positively on the nation's economy as it has enhanced the local engineering capabilities and encouraged the private sector to utilize proven engineering works for raw materials processing. The plants have encouraged the establishment of similar local resource based industries by the private sector. These have raised the supply of locally processed raw materials to Nigerian industries, with a multiplier effect on employment generation, economic empowerment and improved leaving standard of the citizens.

Conscious of the onerous responsibility it has towards the attainment of Nigeria's vision 2020, the Council is strategically repositioned to make meaningful contributions to the transformation of the nation's industrial sector. The organized private sector will constitute the fulcrum of the Council's activities and mandated areas of research, development and utilization through public-private partnership and increased private sector involvement in new resource based industrial ventures.

References

N.S.E., 2009. *Economic Diversification as a Panacea for Sustainable National Development: The Engineers Perspective*, pp. 8–13.

Onwualu, A.P. 2010. *Engineering and Industrial Development*, pp. 18–22.

R.M.R.D.C, 1998. A decade of local sourcing of raw materials (1988-1998), Abuja, PP 30-44.

R.M.R.D.C., 1994. *Investment Profile for Nigeria*, Vol.1 Revised pp. 48–122.

R.M.R.D.C., 2008. *20 Years of Raw Materials Development*, pp. 34–36.

R.M.R.D.C., 2004. *Report on Survey of Agro Raw Materials in Nigeria* (Cocoa), pp. 4–10.

R.M.R.D.C., 1996. *Profile of some Resource-Based Investment Opportunities in Nigeria*, pp. 14–41.

R.M.R.D.C., 2004. *Technical Brief on Minerals in Nigeria (Trona)*, pp. 5–12.

R.M.R.D.C., 2009. *RMRDC Achievements*, pp. 10–23.

Chapter 16

Extraction of Gold from Ore by means of Nanoparticles

Reza Ranjbar, Malek Naderi, Ghasem Amuabedini, Hamid Omidvar* & Mohammad Hassan Mirfassih

Department of Mining and Metallurgical Engineering, Amirkabir University of Technology (Tehran Polytechnic), Iran

**E-mail: omidvar@aut.ac.ir*

ABSTRACT

Anodic Copper slime contains various amounts of precious elements such as Gold, Silver, Palladium, Selenium, etc. (Hironori). All previous research works on Gold extraction by Hydrometallurgical process from ore and Anodic Copper slime have encountered to some extent of problems and difficulties. In this paper extraction of Gold by means of nanoparticles is studied. Magnetite (Fe_3O_4) nanoparticles and Silica-coated Magnetite nanoparticles are used in this study.

***Keywords:** Nanoparticles, Ore, Gold extraction, Copper slime.*

Introduction

In the electro-refining stage of Copper extraction from its ore, impure Copper and pure Copper are used as anode and cathode of the electrolysis cell respectively. Among the impurities only some of them such as Iron, Zinc and Lead are oxidized into ions in the anode. The noble ones such as Gold, Silver, etc remain unchanged and will be poured down at the bottom of the cell gradually as the anode is corroded. These insoluble impurities are called Copper anodic slime (Abdollahi, 1996). There are trace amounts of noble metals in the slime, but their extraction is crucial due to their high value.

Annually, large amounts of Copper anodic slime (about 380 tons yearly) are produced by Iran's Sarcheshmeh Copper complex which contains: Copper, Silver,

Gold, Lead, Selenium, Tellurium, Barium sulfate, etc. The density is also about 4.69 grams per cubic centimeters (Rice *et al.*, 1988).

Many hydrometallurgical methods have been suggested by Subramanian *et al.* Morrison *et al.* Hoffman *et al.* and Rice *et al. which* all of them had problems extracting Gold (Amanian, 1980). In this paper separation and extraction of gold by magnetic nanoparticles is studied.

By definition, all particles having dimensions less than 100 nm are called nanoparticles. Reducing diameter of particles into nano-scale, increases many physical and chemical properties of materials including magnetic, optical and electrical ones (www.nano.ir).

Among nanoparticles, magnetic ones are the most important because of their various applications in Magnetic Resonance Imaging (MRI), disease treatments and separation of materials. Magnetite nanoparticles are the most widely used magnetic nanoparticles due to their good magnetic properties and simple synthesis method (Scherer, 2005).

Magnetite (Fe_3O_4) is one of the most known Iron oxides which is black in color and contains both Fe (II) and Fe (III). Because of its suitable magnetic property, magnetite particles can be easily separated by magnet after their application as sorbents.

Hematite (Fe_2O_3) is another form of Iron oxide. Unlike magnetite, hematite has a low magnetic property. Therefore hematite nanoparticles are not desired in the process. Magnetite particles can be easily oxidized into hematite ones (Laurent *et al.*, 2007).

By coating the surface of the nanoparticles by a narrow layer of Silica (SiO_2), oxidation can be prevented. Although coated nanoparticles have lower magnetic properties, they can still be absorbed by a magnet well (Ferreira, 2009).

Reducing size of particles to nano-scale increases their surface area drastically. Electrical charges on surface of the nanoparticles can make ions in solutions be absorbed. The high surface area makes the ion absorption process more efficient.

There are many ways to synthesize magnetite nanoparticles including co-precipitation, sonochemical, sol-gel and micro-emulsion method, among which co-precipitation is the simplest and the most economical one (Faraji, 2010).

Recently, magnetite nanoparticles are widely used as sorbents of heavy metal ions. In a similar process Gold complex ions inside solution can be absorbed by nanoparticles (www.wikipedia.org/magnetite).

One of the most important steps in hydrometallurgical Gold extraction is leaching. There are various factors that influence the Gold leaching process. For example, the parameters that influence Gold leaching using a mixture of Hydrochloric acid and Nitric acid are:

1. The relative amount of acids,
2. Leaching temperature, and
3. Leaching time.

By changing the parameters and checking the amount of leached Gold in each state, the optimal conditions for leaching can be found out.

Materials and Methods

The Gold inside ore or Copper anodic slime is dissolved in solution. Magnetite nanoparticles are synthesized by the co-precipitation method. In this method $FeCl_2.4H_2O$ and $FeCl_3.6H_2O$ are used as magnetite nanoparticles precursors. The amount of Ammonia solution used in magnetite nanoparticles synthesis is crucial.

To coat the nanoparticles by silica, the Stober process is used. In this process TEOS (Tetraethyl Orthosilicate) is used as silica shell precursor and Ethanol and Ammonium Hydroxide are needed.

To distinguish whether the coating process is successful, Hydrochloric acid is used. Coated nanoparticles are not dissolved in the acid, while nanoparticles without coating will easily be dissolved.

By adding some amount of nanoparticles to Gold solution, Gold complex ions are absorbed on the surface. The absorbed Gold complexes can be desorbed in another solution by changing the pH.

Results and Discussion

The results show that almost all of the Gold inside the solution can be absorbed by magnetite nanoparticles.

The surface of the nanoparticles has a negative electrical charge which can absorb positive charged Gold complexes. Therefore the absorption force is electrostatic.

Conclusions

Magnetite nanoparticles were synthesized successfully. Silica coating was also performed successfully. Gold extraction by magnetite nanoparticles is an efficient process and has a high efficiency.

References

Abdollahi, M., 1996. The treatment of sarcheshmeh copper anode slimes. *PhD. Thesis,* Lids Univ., UK.

Amanian, S., 1980. *Process for the Recovery of Metal Values from Anode Slimes.* US Patent 4229270.

Faraji, M., Yamini, Y. and Rezaee, M., 2010. Magnetic nanoparticles: Synthesis, stabilization, functionalization, characterization and applications. *J. Iran. Chem. Soc.*, 7(1): 1–37.

Ferreira, R.V., Pereira, I.L.S., Cavalcante, L.C.D., Gamarra, L.F., Carneiro, S.M., Amaro, E. Jr., Fabris, J.D., Domingues, R.Z. and Andrade, A.L., 2009. Synthesis and characterization of silica-coated nanoparticles of magnetite. Published online: 26 September 2009 © Springer Science + Business Media B.V.

Hironori Iida. Synthesis of Fe_3O_4 nanoparticles with various sizes and magnetic properties by controlled hydrolysis. *Journal of Colloid and Interface Science.*

Laurent, S. *et al.*, 2007. Magnetic iron oxide nanoparticles: Synthesis, stabilization, vectorization, physicochemical characterization, and biological application" received 2007, pp. 2066–2076.

Rice, N.M. *et al.*, 1988. Charactrization and leaching of copper refinery anode slimes. *LUMA (Leeds Univ. Minig Association) Magazine*, pp. 63–74.

http://www.nano.ir.

Scherer, C., *et al.*, 2005. Ferrofluids: Properties and applications. *Brazilian Journal of Physics*, 35(3A).

www.wikipedia.org/magnetite

Chapter 17

A New Technology in the Surface Coating: Dynamic Chemical Plating (DCP)

Ali M. Taheri, Hamid Omidvar* & Reza Tima

Department of Mining and Metallurgical Engineering,
Amirkabir University of Technology (Tehran Polytechnic), Iran
**E-mail: omidvar@aut.ac.ir*

Introduction

The coatings technologies include films deposited to obtain specific physical properties like optical transparency, where this is combined with a need for resistance to corrosion and wear for example. The value of coatings in the semi-conductor industry, paints applied for purely decorative effect or coatings on jewellery are not reported, coatings that are decorative but have a primary function in protecting the base materials are included (Mathews, 1996).

The present paper will review the fundamentals, technical realization, potentials and advantages of the dynamic chemical plating "DCP" process and gives hints for process selection. Not all details and special variations can be discussed here, but the major aspects influencing the final decision. It will be shown that all processes exhibit advantages and disadvantages – a fact which generally holds for all surface engineering technology. It is worthwhile to briefly compare "Dynamic chemical plating" (DCP) with other deposition techniques.

Category of Processes of Coating

The processes coatings include wet/semi-molten and vapour processing routes (Choy, 2003). Vapour processing routes seem to be the only processing route capable of producing highly pure coatings with structural control at the atomic level or nanometer level and at low processing temperatures, which is important, especially in the microelectronics and opto-electronic industries. There are two well established vapour processing routes, namely Chemical Vapour Deposition (CVD) and Physical Vapour Deposition (PVD). There are hybrid processes such as reactive sputtering and reactive ion plating which introduce O_2, nitrogen or hydrocarbon for the deposition of oxides, nitrides and carbides (or DLC), respectively (Meerakker, 1981; Okinaka and Osaka, 1994). Furthermore, semi-molten processing methods such as plasma spraying and thermal spraying tend to produce porous and splat-like thick coatings, which limit the scope of applications. Wet chemical routes such as slip casting, tape casting, calendaring and screen printing (deposition may be appropriate for thick coating and small area configuration applications, because the large shrinkage associated with the removal of polymeric binders and plasticizers in subsequent sintering steps reduces the film quality for large area fabrication), and electrochemical processes.

Industry Applications

Industry application sectors covered are:

- ☆ Aerospace and medical; aircraft (civil, military and private), aero-engines, missiles and space vehicles, helicopter rotor type vehicles and medical instruments, surgical tools and scientific equipment.
- ☆ Agriculture; does not include produce processing equipment.
- ☆ Automotive; does not include military, aerospace, earth moving or waterborne vehicles.
- ☆ Chemical; plant chemical manufacturing, food processing, brewing, pharmaceutical manufacturing, etc.
- ☆ Construction; domestic and industrial construction materials and structures.
- ☆ Consumer retail; miscellaneous non-electrical products in general, *e.g.* zips, keys, hinges, ladders, etc.
- ☆ Electrical consumer; white (refrigerators, washing machines) and brown (televisions, videos, radios, etc.)
- ☆ Electronic; printed circuit boards and electronic subassemblies.
- ☆ Food and packaging; packing materials used in food, beverage and pharmaceutical industries.
- ☆ Machinery and hand tools; machine tools, drill bits, moulds, wire products, dyes, packaging machinery.
- ☆ Military vehicles; does not include military aircraft or sea-going vessels.
- ☆ Mining; mining or minerals processing and earth moving vehicles.
- ☆ Offshore and marine; fixed constructions used in marine or aqueous environments.
- ☆ Power generation; turbines, power generation and distribution equipment, nuclear power installations.
- ☆ Transport; railway locomotives, rolling stock, track, forklifts, bicycles, conveyors, robots, lifts, escalators, etc.

- ☆ Water and sewage; pipe work for water and effluent, pumps, etc.
- ☆ General; ordinance, small arms, scratch resistant glass coatings, etc.

The Aerosol Chemical Plating

We have described a new metallizing process (Stremsdoerfer, 1997),(Stremsdoerfer, 1997b). It is based on a dynamic concept and that is a complementary technique for the static processes. It combines the advantages of the traditional electroless plating processes and the simplicity of the spray gun techniques. This new process, called by the authors "Dynamic chemical Plating" (DCP), is based on an intermittent oxy-reduction reaction, in an aerosol containing the reducing agent and the metallic salt. The electrochemical reactions occur in a highly dispersed aqueous phase (droplets). The process which avoids the disadvantages listed above and in particularly the Pd/Sn step activation. In few seconds, it is possible to plate a non-conductor surface with metallic films, without a Pd catalytic activation step.

The novelty of this Dynamic Chemical Plating is based on a sequentially highly controlled supply of the material (metallic ions Ni^{+2}, Cu^{+}, Co^{+}, Ag^{+}) and the energy source (reducing agent) to the substrate. In other words it can be described as a metal plating process using intermittent spraying of two or several solutions simultaneously. Using compressed air and a double nozzle spraying gun, the reducing and oxidising agents are sprayed at the same time, but separately, onto a non-conductor or conductor surface (Figure 17.1). Thus, a very thin liquid film is formed. This film is adsorbed on the surface and contains the active species. This film is also thermodynamically unstable which means that the oxidation-reduction reaction will occur spontaneously. But under some experimental conditions, this spontaneous reaction can be controlled to avoid coarse precipitation. These conditions are the following:

- ☆ The sprayed liquid film, which is adsorbed on the substrate, must be thin enough. So the redox reaction occurs on the surface.
- ☆ The ratio of the sprayed oxidizing and reducing solutions must be adequate to give an optimal stoechiometric electrochemical reaction on the surface. This corresponds to the ratio of electrons exchanged between the oxidizing and the reducing agents for a fixed sprayed volume of solutions.
- ☆ The polymer substrate must have a uniform wettability and a relatively high surface energy ($\gamma s > 40$ mJ/m^2). Consequently, a metallic substrate is easier to plate than a plastic material (Stremsdoerfer *et al.*, 2003).

When these conditions are fulfilled, the redox reaction will be induced on the surface instead of a spontaneous precipitation of the metallic ions in the adsorbed liquid film. Then a compact, dense and adherent nickel or copper film is formed on the substrate. Once the reaction is over in the liquid layer, the thin liquid film is refilled with active species by spraying again the solutions.

The time separating two successive refills is called "relaxation time" and during this time the spraying is off. In order to ensure the thickness growth of the metallic film, the spraying is carried out intermittently using successive liquid sprays. The

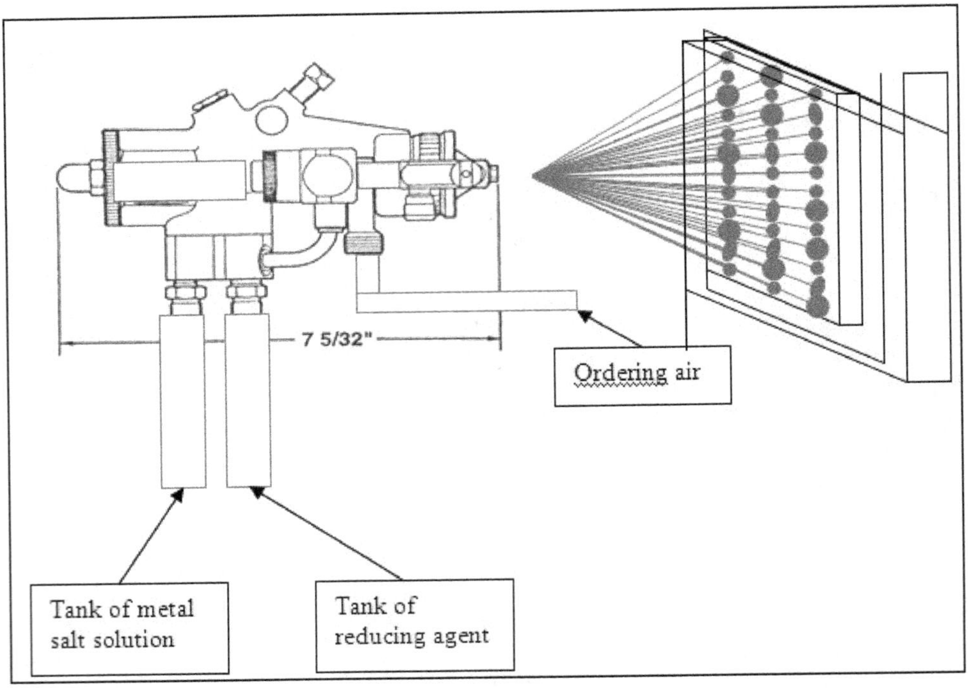

Figure 17.1: Dynamic Chemical Plating Principle

procedure can then be described as a sequence of several spray/non-spray steps or in other words a sequence of several spraying/relaxation steps added. With one of the double nozzle spraying gun, it is possible to cover a surface of 3.8 dm^2.

Advantages of DCP

The dynamic or aerosol chemical plating has many advantages compared to the "static" electroless processes. The absence of a chemical wheel of inertia (chelated metastable solution) avoids the immobilisation and the maintenance of a bath that constitutes a major source of cost in the industry (Stremsdoerfer, 1997). When the parts to be metalized are large-sized, the investment in large dimensions metalizing equipment becomes a real problem. On the other hand, the maintenance and control of the bath stability requires continuous monitoring and analysis effort.

With the new aerosol chemical plating process it is possible, consequently, to intervene on site without being obliged to dismount the parts to be metallized and to treat locally and selectively objects without limitation of size. As this process functions at ambient temperature, it allows an important energy saving compared to the other concurrent techniques which require an operational temperature in the range of 70°C to 90°C. Since the reducing agent is separated from the oxidizing agent there are no stability problems, especially when plating composite coatings (deposits with particles in the metallic matrix).

Results and Discussion

It is worthwhile to briefly compare "Dynamic chemical plating" (DCP) with other deposition techniques such as physical vapor deposition (PVD), chemical vapor deposition (CVD), as well as other non vapor deposition routes such as electroplating and electroless, in order to realize the significance of DCP and their scientific and technological niche markets.

Comparison of Coatings Processes

Physical vapor deposition (PVD) and electrochemical deposition (ECD) both lead to film formation by the deposition of the material in atomic scale.

Table 17.1: Comparison of Typical Coating Parameters for Galvanic, PVD and DCP Processes

Process of Coating	*Dynamic Chemical Plating (DCP)*	*Electroless Plating*	*Electroplating*	*PVD Processes in Batch Mode*
Coating materials	Ni, Ni–B-P, Ni–B, Co, Cu, Ag	Ni, Ni–P, Ni–B, Co, Cu, Pd, Au	Cr, Cu, Ni, Cd, Co, Fe, Sn, Pd, Ag, Rh, Zn and alloys	Metals, alloys, nitrides, oxides, carbides, multilayer
Substrate temperature (°C)	**25-35**	25–100	25–150	100 – evaporation up to 500 – ion plating, sputtering
Process atmosphere	**Room temperature**	Room temperature, chemical reduction	Room temperature, electrolysis	Vacuum, plasma 10-2–10-6 mbar, Ar, reactive gases
Coating thickness (mm)	**0.1 -10**	2–50	2–500 (up to 12 mm)	0.5–50
Coating properties	**To be adjusted**	To be adjusted	To be adjusted, optimized	Must be known in advance, optimized to the substrate
Component size	**No limit**	Limited by bath or barrel	Limited by plating bath	Limited by reactor size
Component material	**No limit**	Cu, brass, iron, steel, Mg, plastics, glass	Easy: steel, Cu, Al, special treatment; stainless steel, Al, Mg, plastics Difficult: Ti, Mo, W, Nb, Ta, glass	No limits, also alloys, plastics, glass

On the base of advanced vacuum and plasma technologies, PVD processes have found their role in coating and thin film deposition, even if not comparable with "traditional" processes regarding the application volume. On the basis of the fundamentals of both techniques the potentials and limitations are outlined. The major advantages of ECD are, for example, the coatability of almost all substrate shapes, specific surface structures, possible leveling of surface roughness or deposition of dispersion coatings. Disadvantages might be the problems with hazardous substances and waste water. The major advantages of PVD are the almost unlimited variation in the chemical composition of the coating material, the principal tolerance of all substrate materials and the deposition of compounds like nitrides, carbides etc. or materials like carbon or diamond. The advantages also are the easy realization of layered or graded film structures. The disadvantages are, for example, the need of vacuum and plasma equipment and the line-of-sight process, with the need for complex sample movement (Hermann, 1999).

1. Difficulty in depositing coatings onto complex shaped components because it is a line-of-sight process;
2. Difficulty for large area deposition and study varying in stoichiometry because it uses solid metal or oxide sources which are tedious and time consuming to be manufactured and installed;
3. Difficulty in controlling the stoichiometry of films (except LASER ablation), which has significant effect on the properties of the films; and
4. Relatively expensive because it needs sophisticated reactor and vacuum systems to facilitate the generation and transportation of vapor species.

Therefore, PVD technologies can easily avoid all the problems of galvanic coatings (wet substrate pre-treatment steps, chemicals, etchants), even for those named 'difficult' in terms of adhesion in Table 17.1. During PVD process efficient low temperature pre-heating and plasma cleaning in the deposition reactor produces excellent adhesion to all substrates. The only limitation is the maximum temperature typical of each PVD process (Navinsek *et al.*, 1999).

CVD-based methods generally involve complex chemistry and chemical process which include the nature of chemical precursors, chemical reactions in the gas phase and heterogeneous reaction. These chemical issues are very often not simple even to chemists. CVD has the following distinctive advantages (Choy, 2003):

1. The capability of producing highly dense and pure materials.
2. Produces uniform films with good reproducibility and adhesion at reasonably high deposition rates. CVD is a non-line-of-sight process with good throwing power. Therefore, it can be used to uniformly coat complex shaped components and deposit films with good conformal coverage. Such distinctive features outweigh the PVD process.
3. It has the ability to control crystal structure, surface morphology and orientation of the CVD products by controlling the CVD process parameters.
4. Deposition rate can be adjusted readily. Low deposition rate is favoured for the growth of epitaxial thin films for microelectronic applications.
5. The flexibility of using a wide range of chemical precursors such as halides, hydrides, organometallics which enable the deposition of a large spectrum of materials including, metal, carbides, nitrides, oxides, sulphides, III–V and II–VI materials.

However, the drawbacks of CVD include:

1. CVD is a complex system.
2. Chemical and safety hazards caused by the use of toxic, corrosive, flammable and/or explosive precursor gases.
3. Difficult to deposit multi-component materials with well controlled stoichiometry using multi-source precursors, because different precursors have different vaporization rates.
4. The use of more sophisticated reactor and/or vacuum system by CVD variants such as low pressure or ultrahigh vacuum CVD, plasma assisted CVD and photo-assisted CVD tends to increase the cost of fabrication.

Conclusions

As a practical application one or other of the techniques will be applied. In other cases the processes might be competing, and often process-combination solves problems. The final decision with respect to the process selection, however, is often governed by non-technical aspects like economic and ecological facts.

In order to reduce the number of electroless plating steps (catalysis step usually carried out in SnCl2/PdCl2 mixed colloidal solutions) and to overcome environmental problems, the metallization of polymers and metals were performed using a direct Dynamic Chemical Plating "DCP" process. The simplicity of "DCP" and its economical aspects make it a competitor process because there is no need for sophisticated apparatus or equipment to perform plating as compared to PVD, CVD. There are no bath precipitation problems, as in the case of static autocatalytic technique. The spraying gun and its components are transportable. DCP is an environmentally friendly process and does not cause any environmental pollution, in fact, there is a full redox reaction on the surface and all the actives species are consumed.

Finally, in few seconds, it is possible to plate a non-conductor or conductor surface with Ni, Cu, Ag, Co films or Ni-B-Graphite, Ni-B-P-Graphite composites and **nano composite** layers at room temperature, without a Pd catalytic activation step for plastics.

References

Choy, K.L., 2003. *Progress in Materials Science*, 48: 57–170.

Hermann, J.A., 1999. *Surface and Coatings Technology*, 112: 210–216.

Karam, A. Fares and Stremsdoerfer, G., 1988. *Plating and Surface Finishing*, 85(1): 88–92.

Mathews, A., Leyland, A. and Stevenson, P., 1996. *Journal of Materials Processing Technology*, 56: 757–764.

Meerakker, J.E.A.M.Van Den, 1981. *Journal of Applied Electrochemistry*, 11: 395.

Navinsek, B., Panjan, P. and Milosev, I., 1999. *Surface and Coatings Technology*, 116–119: 476–548.

Okinaka, Y., and Osaka, T., 1994. Electroless deposition processes: Fundamentals and applications. In: *Advances in Electrochemical Science and Engineering*, Vol. 3, (Eds.) H. Gerisher and C.W. Tobias. VCH Publishers Inc., Weinheim and New York.

Stremsdoerfer and Karam, A. Fares, 1997. *Proceedings of the International Meeting of Electrochemistry*, Ecole Polytechnique de Montréal, Canada.

Stremsdoerfer, G. and Karam, A. Fares, 1997b. French Patent No 2763962.

Stremsdoerfer, G., Ghanem, F., Saikali, Y. and Karam, A. Fares, 2003. *Journal of Materials Science*, 38(15): 3285.

Chapter 18

Gas Nanosensors based on Polypyrrole Nanowires

Nafiseh Hassanzadeh, Hamid Omidvar*, Seied Hadi Tabaian

Department of Mining and Metallurgical Engineering,
Amirkabir University of Technology (Tehran Polytechnic), Iran
**E-mail: omidvar@aut.ac.ir*

ABSTRACT

Gas sensors can detect gas molecules and measure their concentration. Nanostructured materials are attractive for gas-sensing applications owing to their high surface-to-volume ratio, as gas sensors often rely on analyze-surface interactions for their response. Among different nanomaterials for application as sensing elements, conducting polymer nanowires are desirable because of their performance at room temperature. In the current research, polypyrrole nanowires are fabricated by template-synthesis method. For this approach, at first nanoporous alumina templates were fabricated by anodizing of aluminium and after that the nanopores were filled with polypyrrole. By dissolving the alumina template, polypyrrole nanowires which are desirable candidates for using as gas sensing elements, released.

Keywords: *Gas nanosensors, Polypyrrole nanowires, Alumina template.*

Introduction

A sensor is a device that measures a quantity and converts it into a signal which can be read by an observer or by an instrument. Sensors can measure various parameters; physical parameters such as temperature, displacement, acceleration, flow and so on; and chemical and biochemical parameters, such as concentrations of gases, ions or molecules, and molecular interactions can be detected by sensors. The sensor which measures the concentrations of gases is called a gas sensor.

There are innumerable applications for gas sensors in areas including cars, machines, monitoring pollution and environmental conditions, security and safety, aerospace and medicine. Gas sensors can also be useful in industrial plants that use chemicals in manufacturing to detect the release of chemical vapors. When hydrogen fuel cells come into use, in cars or other applications, a sensor that detects escaped hydrogen could be very useful in warning of a leak.

Many applications demand miniaturisation to reduce power consumption for integration into portable devices. Affordable mass production is also a prerequisite for sensors for consumer products, and for disposable devices such as sensors for pollution monitoring.

Technological progress allows more and more sensors to be manufactured on a microscopic scale as microsensors. In most cases, a microsensor reaches a significantly higher speed and sensitivity compared with macroscopic approaches.

Nanotechnology Impact on Sensors

Nanostructured materials are attractive for gas-sensing applications owing to their high surface-to-volume ratio, as gas sensors often rely on analyte-surface interactions for their response.

Nanotechnology offers a unique advantage to the sensor industry by manipulating materials at the molecular level, where sensing events occur. Potentially,

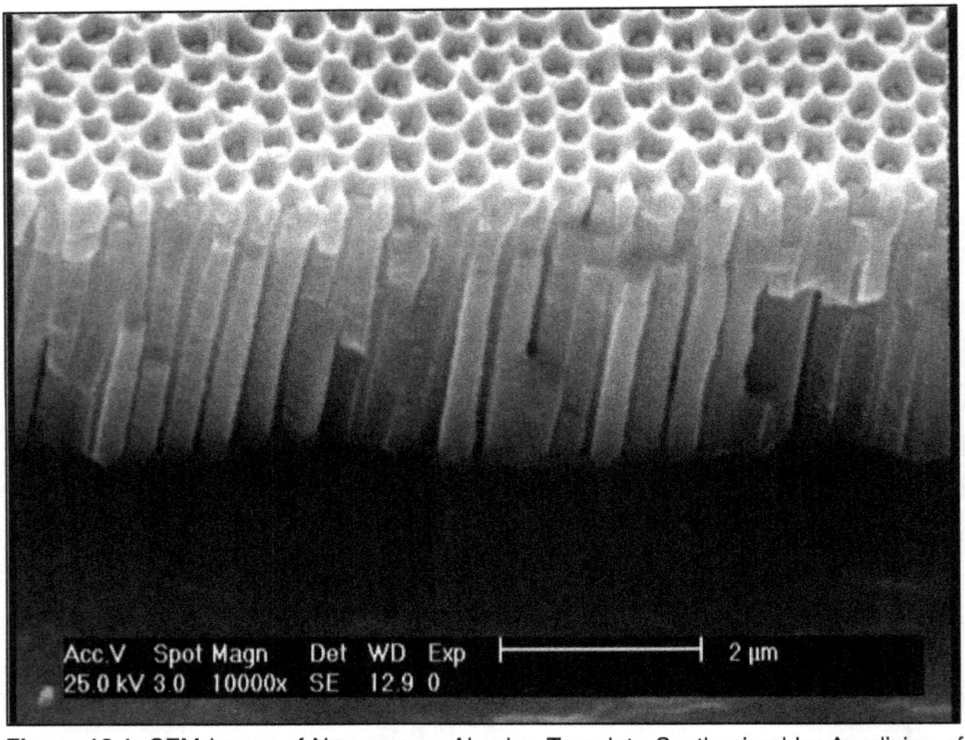

Figure 18.1: SEM Image of Nanoporous Alumina Template Synthesized by Anodizing of Aluminium

the technology offers devices with greater sensitivity, selectivity, and reliability than conventional sensors. Various types of detecting elements, such as carbon nanotubes, metal oxides, conducting polymers or palladium nanostructures can be used in nanotechnology-based sensors. These detecting elements change their electrical characteristics, such as resistance or capacitance, when they absorb a gas molecule.

Because of the high surface-to-volume ratio of nanotubes, nanowires, or nanoparticles, a few gas molecules are sufficient to change the electrical properties of the sensing elements. This allows the detection of a very low concentration of chemical vapours.

Nanotechnology is certain to improve existing sensors and be a strong force in developing new ones and the goal is to have small and inexpensive sensors. In the report by Andrew McWilliams, nanosensors represented 10 per cent of the total sensors market in 2010 and the global nanosensors market reaches 12 billion by 2012.

Polypyrrole Nanowires as Sensing Elements

Polypyrrole nanowires are desirable candidates for application as sensing elements, because they can detect gas molecules at room temperature while metal oxide gas sensors work at high temperatures above 300°C.

In this research, in order to fabricate polypyrrole nanowires, "template-synthesis" method has been used. For this approach, at first, alumina templates have been

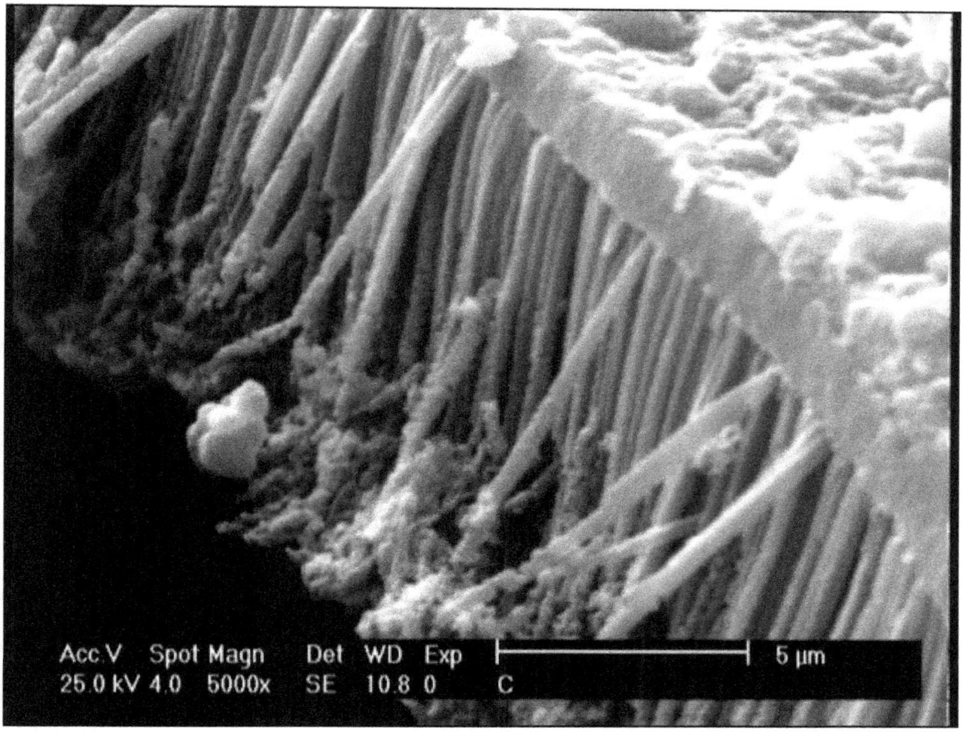

Figure 18.2: SEM Image of Polypyrrole Nanowires after dissolving the Alumina Template

fabricated by anodizing of aluminium. Anodizing process was done in 0.3 M oxalic acid electrolyte under constant voltage of 40 V at 17 °C. The produced nanostructure consists of closed-packed array of hexagonally arranged cells containing pores in each cell-center. The SEM image of this nanoporous film is shown in Figure 18.1.

The uniform pore diameter, which is easily controllable by altering the anodizing conditions, can range from 50 nm to 200 nm.

After the synthesis of alumina template, the nanopores of the template were filled by polypyrrole. For this approach, a two-compartment cell was designed and an aqueous pyrrole solution (0.2 M) and the oxidant reagent ($FeCl_3$, 0.2 M) was introduced in each cell. The monomer and the oxidant reagent diffuse toward each other through the pores of the membrane and react to yield the polymer. After the synthesis, alumina template was dissolved and conducting polypyrrole nanowires released. The SEM image of these nanowires is shown in Figure 18.2.

By designing the electrical circuits, the changes of electrical conductivity of polypyrrole nanowires after detecting gas molecules can be observed and recorded.

References

Waghuley, S.A. *et al.*, 2008. *Sensors and Actuators,* B128: 366–373.

Xing-Jiu Huang *et al.*, 2007. *Sensors and Actuators,* B122: 659–671.

Chapter 19

IROST Green Technology Pilot Plant: Microalgae Based Fuel

Nasrin Moazami*[1], Reza Ranjbar[2], Alireza Ashori[3], Mehrnoush Tangestani[1] & Roqieh Eghtesadi Ali Sheykhi Nejad[1]

[1]*Department of Biotechnology, Iranian Research Organization for Science and Technology (IROST), Tehran, Iran*

[2]*Faculty of Chemical Engineering, Amirkabir University of Technology (AUT), Tehran, Iran*

[3]*Department of Chemical Technologies, Iranian Research Organization for Science and Technology (IROST), Tehran, Iran*

**E-mail: moazami@irost.org*

ABSTRACT

Research on microalgae and cyanobacteria in IROST begain in 2002 when we established the Persian Gulf Biotechnology Research Center in Qeshm Island. In 2008 with financial support of Ministry of Industries and Mines, we carried out a project in pilot scale for production of Biodiesel, Bioethanol, and high value products from our strain of microalgae selected from Persian Gulf and Qeshm Island coastline. Eight hundred marine water samples from Persian Gulf were transferred to the laboratory. 147 microalgae strain were isolated and screened for their biomass productivity, lipid content and other valuable products. Twenty four strains, selected because they were robust, highly productive and with a relatively high lipid content. The strain PTCC 6016 which had 52 per cent lipid was grown in 200 liters open pond photo bioreactor to study the influence of culture conditions. The results were used to design the IROST Green Technology Pilot Plant.

Keywords: *Microalgae, Biofuel, Biodiesel, Biomass, Lipid, Open pond.*

Introduction

One of the most serious environmental problems is the global warming caused primarily by the heavy use of fossil fuel. Sustainability and environmental concern are major issues in the world today. It is now a widely accepted necessity to reduce carbon dioxide emissions from human activities contributing to the greenhouse effect and thus global warming. The use of resources from agricultural sector to produce biofuel generally induces a lower climate change potential, but can create other environmental issues and generate competition with food crops for the use of arable land.

Algae use as food has been cited in Chinese literature as early as 2500 years ago (Tseng, 2004). Several parts of Asia and some indigenous people in Africa, South America and Mexico consume naturally occurring algae mostly because of the vitamins and nutrients they provide (Edwards, 2008). Due to the fact that algae are very efficient at converting light, water and carbon dioxide into biomass in a system that does not necessary require agricultural land, algae are attractive as a sustainable alternative because they do not compete with agricultural food and freshwater (FAO, 2009a). Microalgae grow very fast compared to terrestrial crops, they commonly double in size every 3-6 hours. Oil content of microalgae is usually between 20 to 50 per cent of dry weight, while some strain can reach as high as 70 per cent (Chisti, 2007; Pulz and Gross, 2004). Table 19.1 show yield based on crop type and microalgae.

Table 19.1: Oil Yield Based on Crop Type and Microalgae

Crop	*Oil Yield (Gallons/acre)*
Corn	18
Soybeans	48
Canola	127
Jatropha	202
Coconut	287
Oil Palm	636
Microalgae[1]	6283–14641

Source: Adapted from Chisti, 2007

1 Oil content ranges from 30 per cent to 70 per cent of dry biomass.

Through photosynthetic metabolism, microalgae absorb CO_2 and releases oxygen. On a global scale microalgae produce more than 75 per cent of the oxygen required for living of planet Earth.

It is also widely accepted that except for algae, no other bio-feedstock have the ability to replace the entire global fossil fuel requirements. Possible algal biofuels include biodiesel, bio-ethanol, biocrude, biogas, jet-fuel and biohydrogen (Amin, 2009). While important non fuel options include the protein, pigments, essential oil and other bioactive compounds as health food, nutraceuticals, cosmeceuticals, pharmaceuticals or other renewable inputs for the food industries, including feed for livestock, and aquaculture (Lee, 1997; Chisti, 2007; Spolaore *et al.*, 2006; Mata *et al.*, 2010). Our objective of this research was to maximize biomass and lipid productivities and design the IROST pilot plant for large scale biodiesel production.

Materials and Methods

Screening of Microalgae with High Biomass Productivity in the Period of Two Weeks

800 samples of Persian Gulf were transferred to laboratory and cultivated in different culture media to isolate and purify the strains. To select the algae with high productivity micro algae was maintained in sea water enriched by 7.5 per cent $NaNO_3$ and 0.5 per cent NaH_2PO_4 in 5 liter photo-bioreactor at 28°C, under 14 h. illumination (130 μmol PAR photon/m²/s) provided by florescent light tube. Paddle well maintained with 20 rpm, CO_2 was added to pond to provide carbon source to microalgae and maintained pH to 7.5 in first week of cultivation. In the second week with the rise of pH to over 8, microalgae concentrated high quantity of oil. CO_2 was injected into the culture through a gas diffuser for pH regulation. Each strain was cultivated in single batch for two weeks (Figures 19.1 and 19.2).

Wet algae biomass was harvested everyday and the cell density was measured (Unicom UV.Vis spectrometry) using 590 nm wavelength. Productivity was calculated by optical density.

Screening of Microalgae with High Oil Yield from Rapid Growth Microalgae

Selected micro algal cells were tested for Nile red staining. Nile red (9-(Diethylamino)-5H-benzo[a]phenoxazin-5-one) staining was conducted to detect intracellular lipid droplets (Rodolfi *et al.*, 2008). The cultured cells (0.5 ml) were collected by centrifugation for 10 min and washed with physiological saline solution (0.5 ml) several times. After the collected cells were re-suspended in the same solution (0.5 ml), the Nile red solution (0.1mgmL^{-1} in acetone) was added to cell suspensions (1:100 v/v) and incubated for 10 min. After washing once, stained microalgae cells were observed by fluorescent microscopy (Figure 19.3).

Optimization of Growth Conditions in 200 liters Open Pond

Three of the most promising strains were cultivated in 200 liter open pond to investigate biomass productivity and lipid synthesis (Brennen and Owende, 2010). The open pond containing 200 liter media, were mixed with paddle well. CO_2 injected to support growth and maintain pH within the range of 7.5- 8.5. Continuous artificial illumination of 180 μmol PAR photons/m²/s was provided by daylight fluorescent tube on the top of open pond. A batch and continuous regimen of 25 per cent culture carried out to evaluate the different culture condition (Figure 19.4).

Separation of Biomass

Wet algae biomass was harvested everyday and the cell density was measured (Unicom UV.Vis spectrometry) using 590 nm wavelength. Productivity was calculated by optical density after two weeks algal biomass was transfer to a collection tank, flocculate biomass dries for 48 h. to approximately 90 per cent solids content.

Dried biomass was initially ground in a Wiley mill to pass 3 mm. sieves and stored in sealed plastic bag at 20-25°C. The growth curve was followed by measuring

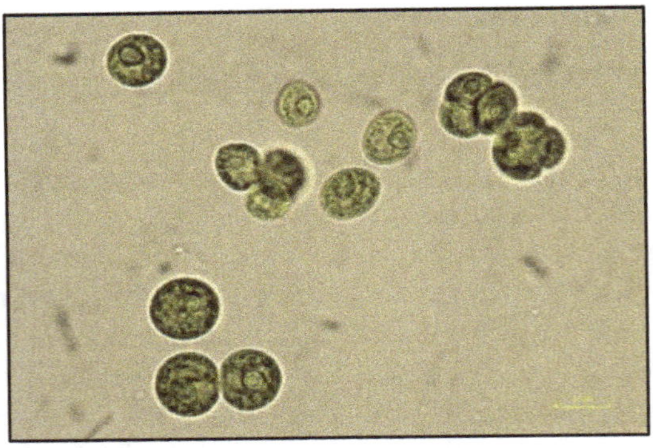

Figure 19.1: Early Stage of Cultivation

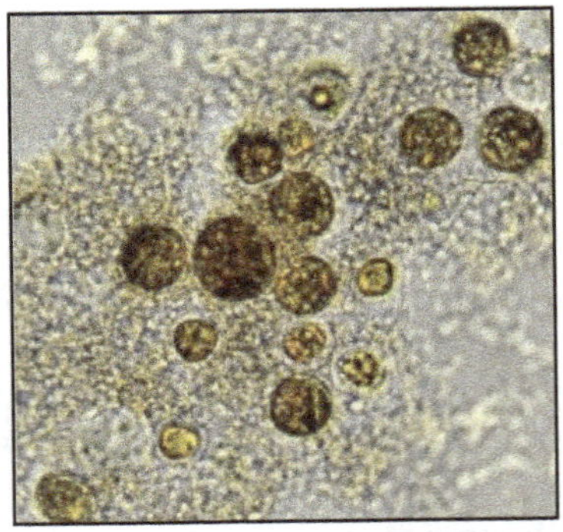

Figure 19.2: Oil Concentration in the Cells

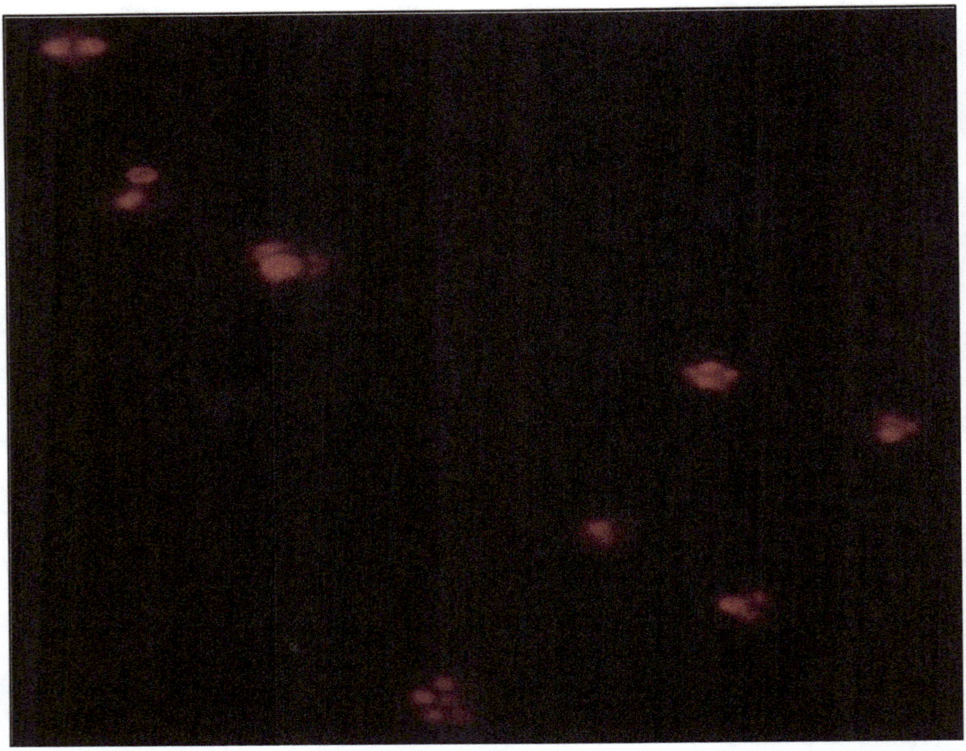

Figure 19.3: Nile Red Staining of Fatty Acid in Microalgae

the optical density at 590 nm. Daily biomass productivity was calculated dividing the difference between the day weight at the start and at the end of the experiment by its duration.

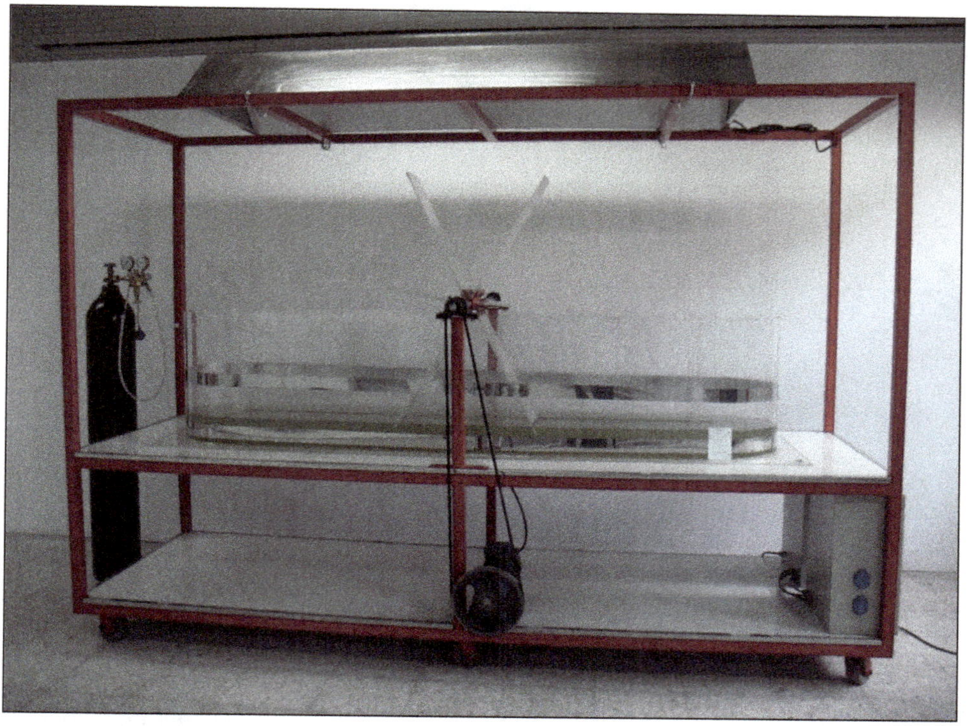

Figure 19.4: 200 liter Photo-Bioreactor

Oil Extraction

Oil content was extracted from algae biomass, using a modified Bligh and Dyer method (1959). Dried algae were extracted with chloroform/methanol (2:1 v/v) at 25°C at room temperature to allow the organic and aqueous layers to separate. After removing oil layer and evaporation of solvent, oil esterifies and was analysed for Fatty Acid composition using gas chromatography (GC) method.

Modeling and Design of the Indoor and Outdoor Open Pond

Research focused on the design of pilot plant in particular on the efficiency of light energy conversion into algal biomass in open pond reactor (Rosenberg *et al.*, 2008) using" *Tracpro" soft ware.* These are indoor (greenhouse) and outdoor open systems in which algae can be cultivated with sunlight as the energy source at a high degree of control and optimization. In addition design focused on methods to manipulate culture condition in such a manner that the algae started to produce commercial products such as antioxidants and essential fatty acids in excess. The algae research and production center is commonly known as "IROST Green Technology Pilot plant".

Results and Discussion

The exponential phases are reached within four days and lasts about 10 days, among 147 strains, 24 strains of screened algae were selected.

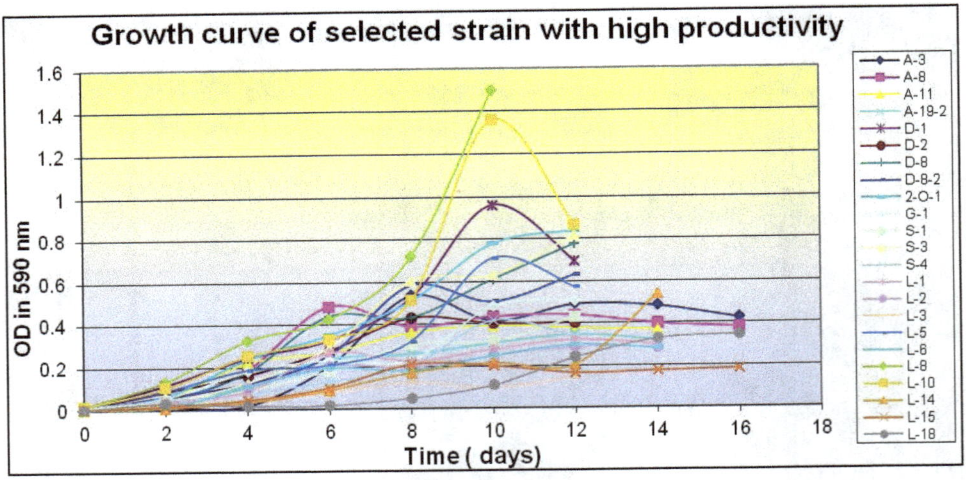

Figure 19.5: Growth Curve of 24 Selected Strains

It has a high growth rate, which enhance reaching high yield biomass and lipid per unit volume of broth (Figure 19.5). The bioreactor allows approximately a yield of 2 gm dry microalgae biomass per liter of algae broth. The Nannochloropsis PTCC 6016 with a lipid content of 52 per cent and lipid productivity ranging 55 to 70 mg/L/day, showing the best combination of biomass productivity and lipid content, perform well in batch cultivation. Figure 19.6 shows the lipid productivity of PTCC 6016. (Moazami *et al.*, 2011).

These microalgae were used for pilot production of biodiesel (Figure 19.7). The pilot plant and test facility is just a start of what should became the leading algal research center in IROST, particularly research and development of methods aimed at making specific end products will be further expanded in the coming years. IROST

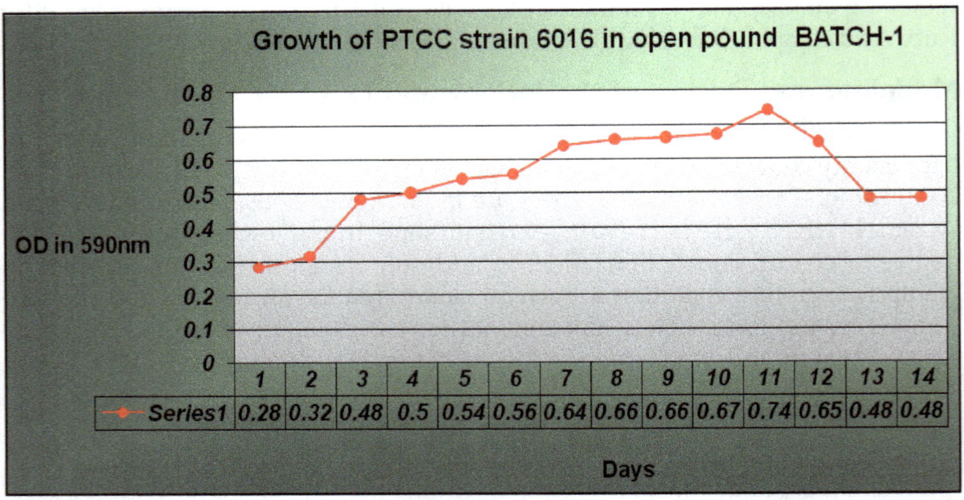

Figure 19.6: Batch Cultivation of PTCC 6016 Strain in 200 litre Reactor

Figure 19.7: Microalgae Biodiesel

Green Technology pilot plant consist of six, 2000 liter indoor (greenhouse) open pond, two 2000 liter and three 25,000 litre outdoor open pond. The reactor contained paddle well for mixing and gas exchange. Paddle well was maintained with 20 rpm/min. A control unit provided optimum pH, temperature, water level and evaporation records during the two weeks experiments. The cooling system in indoor open pond prevented the culture temperature to exceed the value of 30°C. During the night, the culture temperature was allowed to equilibrate to ambient. Figures 19.8 and 19.9 show the indoor and outdoor Open pond of IROST Green Technology Pilot Plant.

Figure 19.8: Indoor Open Pond

Figure 19.9: Indoor and Outdoor Open Pond of IROST Green Technology Pilot Plant

Conclusion

With depletion of petroleum based fuel sources, rising crude oil and gas prices, and global warming related to use of fissile fuels, domestics production of energy from renewable resources could become attractive. Algae have the potential to be a significant source for biofuels. Current biodiesel production from oil seed crops face a dilemma in that they compete with food supplies. By producing algae biodiesel, farmland can be reserved for food production (Pienkos and Darzin, 2009). Establishment of IROST Green Technology pilot plant play a key role in the training and research facility for optimization of algae cultivation. The research team tested various cultivation systems. Based on these results and data obtained from the pilot, the team will develop the reactor design for application on commercial scale.

Acknowledgment

The Ministry of Industries and Mines of Iran is gratefully acknowledged for providing financial assistance to carry out this research work.

References

Amin, S., 2009. Review on biofuel oil and gas production processes from microalgae. *Energy Conversion and Management*, 50: 1834–1840.

Bligh, E.G. and W.J. Dyer, A., 1959. A rapid methods of total extraction and purification. *Canadian Journal of Biochemistry and Phisiology*, 37(8): 911–917.

Brennan, L. and Owende, P., 2010. "Biofuels from microalgae: A review of technologies for production, processing, and extractions of biofuels and coproducts." *Renewable and Sustainable Energy Reviews*, 14(2): 557–577.

Chisti, Y., 2007. Biodiesel from microalgae. *Biotechnol. Adv.*, 25: 294–306.

Demirbas, A., 2010. Use of algae as biofuel sources. *Energy Convers. Manage.*, 51: 2738–2749.

Edwards, M., 2008. *Green Algae Strategy End Biowar and Engineer Sustainable Food and Biofuels*. Tempe, Arizona, USA, LuLu Press.

FAO, 2009a. *Algae-Based Biofuels: A Review of Challenges and Opportunities for Developing Countries*. FAO, Rome, Italy.

Lee, Y.K., 1997. Commercial production of microalgae in the Asia-Pacific rim. *Journal of Applied Phycology*, 9(5): 403–411.

Mata, T.M. and Martins, A.A. *et al.*, 2010. Microalgae for biodiesel production and other applications: A review. *Renewable and Sustainable Energy Reviews*, 14(1): 217–232.

Moazami, N., Reza R., Alireza A., Mehrnoush T. and Ali S.N., 2011. Biomass and lipid productivities of marine microalgae isolated from the Persian Gulf and the Qeshm Island. *Biomass and Bioenergy*, 35: 1935–1939.

Pienkos, P.T. and Darzins, A., 2009. The promise and challenges of microalgae derived biofuel. *Biofuels, Bioproducts and Biorefinering*, 3(4): 431–440.

Pulz, O. and Gross, W., 2004. Valuable products from biotechnology of microalgae. *Applied Microbiology and Biotechnology*, 65(6): 635–648.

Richmond, A., 2004. *Handbook of Microalgal Culture: Biotechnology and Applied Phycology*. Blackwell Science, Oxford [etc.].

Rodolfi, L., Zitteli, G.C., Bassi, N., Padovani, G., Biondi, N., Bionini, G. and Tredici, M.R., 2008. Microalgae for oil: Strain selection, induction of lipid synthesis and outdoor mass cultivation in a low-cost photobioreactor. *Biotechnology and Bioengineering*, 102(1): 100–112.

Rosenberg, J.N. and Oyler, G.A. *et al.*, 2008. A green light for engineered algae: Redirecting metabolism to fuel a biotechnology revolution. *Current Opinion in Biotechnology*, 19(5): 430–436.

Spolaore, P., Joannis-Cassan, C., Durass, E. and Isambert, A., 2006. Commercial application of microalgae. *Journal of Bioscience and Bioengineering*, 101(2): 87–96.

Tseng, C.K., 2004. The past, present and future of phycology in China. *Hydrobiologia*, 512(1–3): 11–20.

www.ingramcontent.com/pod-product-compliance
Lightning Source LLC
Chambersburg PA
CBHW070250260926
51849CB00001BA/92
* 9 7 8 9 3 5 1 2 4 1 7 4 4 *